Mirko Holler

Attosecond Strong Field Control

Mirko Holler

Attosecond Strong Field Control

Control of quantum path interferences in high harmonic generation and the absorption of attosecond pulse trains in helium

Südwestdeutscher Verlag für Hochschulschriften

Impressum/Imprint (nur für Deutschland/ only for Germany)
Bibliografische Information der Deutschen Nationalbibliothek: Die Deutsche Nationalbibliothek verzeichnet diese Publikation in der Deutschen Nationalbibliografie; detaillierte bibliografische Daten sind im Internet über http://dnb.d-nb.de abrufbar.
 Alle in diesem Buch genannten Marken und Produktnamen unterliegen warenzeichen-, marken- oder patentrechtlichem Schutz bzw. sind Warenzeichen oder eingetragene Warenzeichen der jeweiligen Inhaber. Die Wiedergabe von Marken, Produktnamen, Gebrauchsnamen, Handelsnamen, Warenbezeichnungen u.s.w. in diesem Werk berechtigt auch ohne besondere Kennzeichnung nicht zu der Annahme, dass solche Namen im Sinne der Warenzeichen- und Markenschutzgesetzgebung als frei zu betrachten wären und daher von jedermann benutzt werden dürften.

Verlag: Südwestdeutscher Verlag für Hochschulschriften Aktiengesellschaft & Co. KG
Dudweiler Landstr. 99, 66123 Saarbrücken, Deutschland
Telefon +49 681 37 20 271-1, Telefax +49 681 37 20 271-0
Email: info@svh-verlag.de
Zugl.: Zürich, ETH Zürich, Diss. , 2010

Herstellung in Deutschland:
Schaltungsdienst Lange o.H.G., Berlin
Books on Demand GmbH, Norderstedt
Reha GmbH, Saarbrücken
Amazon Distribution GmbH, Leipzig
ISBN: 978-3-8381-1851-2

Imprint (only for USA, GB)
Bibliographic information published by the Deutsche Nationalbibliothek: The Deutsche Nationalbibliothek lists this publication in the Deutsche Nationalbibliografie; detailed bibliographic data are available in the Internet at http://dnb.d-nb.de.
 Any brand names and product names mentioned in this book are subject to trademark, brand or patent protection and are trademarks or registered trademarks of their respective holders. The use of brand names, product names, common names, trade names, product descriptions etc. even without a particular marking in this works is in no way to be construed to mean that such names may be regarded as unrestricted in respect of trademark and brand protection legislation and could thus be used by anyone.

Publisher: Südwestdeutscher Verlag für Hochschulschriften Aktiengesellschaft & Co. KG
Dudweiler Landstr. 99, 66123 Saarbrücken, Germany
Phone +49 681 37 20 271-1, Fax +49 681 37 20 271-0
Email: info@svh-verlag.de

Printed in the U.S.A.
Printed in the U.K. by (see last page)
ISBN: 978-3-8381-1851-2

Copyright © 2010 by the author and Südwestdeutscher Verlag für Hochschulschriften Aktiengesellschaft & Co. KG and licensors
All rights reserved. Saarbrücken 2010

Table of Contents

Table of Contents I
List of Figures III
Publications IX
    Journal Papers IX
    Conference Papers X
Abstract XVI
Kurzfassung (German) XVIII
Introduction 1

High Harmonic Generation 6
    2.1   High harmonic generation 6
        2.1.1   Single atom response 8
        2.1.2   Electron trajectories and intrinsic harmonic phase 10
        2.1.3   Quantum mechanical formulation 14
        2.1.4   Macroscopic response 17
        2.1.5   Temporal structure of high harmonics 19
    2.2   Laser systems 20
        2.2.1   Ti:Sapphire laser 21
        2.2.2   The laser system used for the experiments 23

Quantum Path Interference 26
    3.1   Theoretical description of QPI 26
    3.2   Experimental conditions for the observation of QPI 28
        3.2.1   Temporal averaging 28
        3.2.2   Spatial averaging 30
    3.3   Experimental setup 32
    3.4   Experimental results 33
        3.4.1   Argon 37
        3.4.2   Xenon 40
        3.4.3   Neon 41
    3.5   Influence of CEP, phase-matching and phase modulation on QPI 43
    3.6   Spatially resolved QPI 43
    3.7   Conclusion 45

A New Beamline For Attoscience 47
    4.1   Basic requirements and optical layout 48
    4.2   Overview of the vacuum system 52
    4.3   Generation chambers 55
        4.3.1   Motorized optical components 58
        4.3.2   Gas targets for high harmonic generation 59
            4.3.2.1   The pulsed valve 59
            4.3.2.2   The capillary target 61
        4.3.3   Filter wheel for metallic thin-film foils 62
    4.4   Diagnostics and refocusing chamber 63
    4.5   Interaction chamber 66
    4.6   Refocusing and XUV spectrometer 68

4.7 Active beam stabilization 69

Characterization of APTs 71
5.1 A beam profile of high harmonic radiation 71
5.2 Properties of thin-film filters, measurement of the transmission of an aluminum filter foil 72
5.3 Measurement of the flux of high harmonics 75
5.4 Temporal characterization of an APT 77
    5.4.1 Theoretical background of RABBITT 78
    5.4.2 Experimental setup 80
    5.4.3 Experimental results 84

Attosecond Transient Absorption 88
6.1 Experimental setup 88
6.2 Absorption of high harmonics for large delays 90
6.3 Typical delay scan 93
6.4 Femtosecond structure 94
6.5 Attosecond structure 95
    6.5.1 Phase differences in the modulation of the transmission of individual harmonics 95
    6.5.2 Total transmitted photon yield 98
6.6 Asymmetric structure in the absorption spectra of harmonic 15 99
6.7 Theoretical description of the experiment 101
6.8 Conclusion and outlook 102

Summary And Outlook 103

References 107

Appendix 119

List of Abbreviations 121

Curriculum Vitæ 123

Danksagung 125

List of Figures

Fig. 2.1 Typical high harmonic spectrum generated by a multi-cycle driving field showing the plateau with nearly constant intensity, followed by the cut-off..........7

Fig. 2.2 Regimes of atomic ionization by low energetic photons ($E_{photon} < I_p$), where I_p is the ionization potential of the atom. 8

Fig. 2.3 Illustration of the three-step-model [12]. See text for details....................9

Fig. 2.4 Electric field (blue) and position of the electron for different electron-release-phases versus time (arb. u.). Depending on the release phase there are trajectories that return to the ion (green), and trajectories that do not return (red). 11

Fig. 2.5 Classical electron trajectories and their energy upon return to the nucleus. One cutoff trajectory can be identified. For lower energies two trajectories exist, one with a long and one with a short excursion time. 12

Fig. 2.6 Energy-spectrum of the returning electron versus phase of electron-release (red). The black line shows the time the electron spent in continuum (arb. u.). 13

Fig. 2.7 A multi-cycle driving pulse will generate a train of attosecond pulses in time domain. The separation of the pulses is one half-cycle of the driving field. 20

Fig. 2.8 Schematic of the laser system used in the experiments presented in this thesis. PC = Pockels cell, D = Dazzler. 25

Fig. 3.1 The intrinsic harmonic chirp (blue, arb. u.) spectrally shifts the harmonic emission generated at different instants in the laser pulse (red, arb. u.). The chirp is different for the long and the short trajectory (see also [63]). The central frequency of the harmonic emission is for both trajectories generated at the peak intensity of the laser pulse. 29

Fig. 3.2 Calculated macroscopic spectra of the 15th harmonic generated in argon as a function of laser peak intensity without (left) and with (right) a far-field 6-mrad off-axis window. The harmonic signal is plotted with a linear color scale defined as blue at zero strength to red at maximum strength (10^{10} without window and $4.5 \cdot 10^7$ with window). By courtesy of T. Auguste. 31

Fig. 3.3 Schematic experimental setup. A detailed description is given in the text. ...32

Fig. 3.4 High harmonic beamline used to perform the experiments on QPI. 33

Fig. 3.5 Spectral intensity of high harmonics versus laser peak intensity generated in argon. The jet has been placed after the laser focus, preferably selecting the short trajectory in the phase matching process. In the upper image, the spatial selection was positioned on-axis, on the lower image it was off-axis. 35

Fig. 3.6 Spectral intensity of high harmonics versus laser peak intensity generated in argon. The jet has been placed before the laser focus, the short and the long trajectory are both phase matched. In the upper image, the spatial selection was positioned on-axis, on the lower image it was off-axis. Compared to the on-axis selection, the long trajectory with its larger divergence is detected in the case of an off-axis selection which leads to a spectral broadening of the individual harmonics. 36

Fig. 3.7 Harmonic spectra generated in argon versus laser peak intensity with short and long trajectory phase-matched and off-axis spatial selection:

(a) measured spectra, (b) simulated spectra around harmonic 21, courtesy of
T. Auguste. 37

Fig. 3.8 Spectrally integrated signal in a narrow window at harmonic order 21
from Fig. 3.7 (a) showing several modulations of the harmonic yield due to
QPI. 38

Fig. 3.9 Simulated QPI for H15 in argon at a pressure of 10 Torr. (a) full
simulation, (b) ionization switched off, (c) with depletion, but free electron
dispersion switched off. By courtesy of T. Auguste. 39

Fig. 3.10 (a) Measured harmonic spectra generated in xenon versus laser peak
intensity, (b) simulated spectrum for harmonic order 17, courtesy of
T. Auguste. 40

Fig. 3.11 Spectrally integrated signal in a narrow window at harmonic order 17
from Fig. 3.10 (a) showing the modulation due to QPI. Only one period can be
observed due to fast depletion of the medium. 41

Fig. 3.12 (a) Experimental harmonic spectra generated in neon with respect to the
laser peak intensity: over the full range of accessible intensities located below
ionization threshold neither a blue shift nor a saturation of the harmonic signal
at the harmonic central frequency is observed, (b) simulated spectrum for
harmonic order 35, courtesy of T. Auguste. 42

Fig. 3.13 Spectrally integrated signal in a narrow window at harmonic order 35
from Fig. 3.12 (a) showing the modulation due to QPI with a modulation
periodicity of ~ $8 \cdot 10^{13}$ W/cm². 42

Fig. 3.14 Scheme of the experimental setup for the observation of spatially
resolved QPI. The strong emission of the short trajectory is indicated in dark
blue. It is surrounded by its weaker part plus the emission from the long
trajectory. The beam was misaligned on the entrance slit to balance the
contributions from the two trajectories. This enabled the observation of
spatially resolved QPI. 44

Fig. 3.15 Spatially resolved interference structure measured at a fixed peak
intensity of $3.4 \cdot 10^{14}$ W cm⁻² for harmonic order 19 and 21 generated in argon. ..45

Fig. 4.1 Typical XUV/IR pump-probe measurement shown for two different time
delays τ_1 and τ_2. The attosecond pulses (labeled with XUV) are synchronized
to the driving IR field (IR). 48

Fig. 4.2 Basic optical layout of the new high harmonic beamline. 49

Fig. 4.3 Reflectivity of a gold surface at an angle of incidence of 82 degrees for p-
polarized light in the typical wavelength range of the high harmonics [71]. 50

Fig. 4.4 Raytracing performed using ShadowVUI [72]. Generation focal spot
(beam diameter 60 μm, left); interaction focal spot (beam diameter 60.6 μm,
center); beam profile on the entrance slit of the spectrometer (beam height
1.94 mm, beam width 77 μm, right). 52

Fig. 4.5 Damper for turbomolecular pumps, reducing the vibrations transferred to
the setup by 25 dB (at the rotation frequency of the turbopump). The two
flanges are only connected via a flexible bellow which is stabilized by rubber
dampers. 53

Fig. 4.6 Overview of the harmonic beamline and its vacuum chambers. (ABS =
active beam stabilization) 54

Fig. 4.7 Schematic of the three identical generation chambers on the base plate.
The chambers are connected via rectangular flanges. Between the second and
third chamber a differential pumping stage is indicated. 55

LIST OF FIGURES

Fig. 4.8 Sealing screw for fixing the aluminum breadboard to the optical table from inside vacuum. 56

Fig. 4.9 Section of a part of the generation chambers. The holes that allow for fixing the breadboard to the optical table using the sealing screws are visible (Fig. 4.8). 57

Fig. 4.10 Mechanical stability measured by capturing interferences in an optical interferometer using a Helium Neon laser and a CCD camera (A501k, Basler AG). At two instants (27 s and 32 s) mechanical shocks have been produced on the optical table. The interferometer recovers to its initial position. 58

Fig. 4.11 Images of the pulsed valve. On the left the valve is shown with the *xyz* translation stage. On the right a section of the valve is shown. The red part is a piezoelectric disc which is operated synchronously to the repetition rate of the laser. It moves a pin with an O-ring that seals the valve when no voltage is applied to the piezo. 60

Fig. 4.12 Sections of two target geometries for the pulsed gas jet. The open target (left) offers a short interaction length, the "T"-shaped target (right) offers a high gas density. 60

Fig. 4.13 Sectioning of the capillary target. The laser beam enters from left through the entrance window and is focused into the capillary. The capillary forms the gas outlet to vacuum. The enclosed volume is continuously filled with gas for HHG. An observation window allows for visual inspection of the capillary entrance. 62

Fig. 4.14 Exploded view of the filter wheel. 63

Fig. 4.15 Schematic of the calibrated photodiode from NIST (National Institute of Standards and Technology, USA) and its quantum efficiency (electrons per photon). 64

Fig. 4.16 View into the diagnostics and refocusing chamber. The beam enters from the right. 65

Fig. 4.17 Section of the interaction chamber with the long time-of-flight tube attached. The laser beam enters from the bottom. The TOF tube is made from stainless steel with a 1 mm thick mu-metal tube inside to prevent magnetic fields from influencing the propagation of the electrons. 67

Fig. 4.18 The entrance of the time-of-flight tube. The gas target is shown in red. It is moved into the mu-metal shielding of the TOF-tube. 67

Fig. 4.19 Spectrometer unit with refocusing chamber and the commercial XUV spectrometer. 69

Fig. 5.1 Typical beam profile of high harmonics. On the left the CCD image is shown. The right graph shows an according line profile (red) with a Gaussian fit (blue). 72

Fig. 5.2 Transmission of thin-film foils for the case of aluminum and zirconium, each with a thickness of 100 nm [71]. 72

Fig. 5.3 Left: Two harmonic spectra generated in argon with identical conditions. One spectrum (A_1) is recorded with one filter (shown in blue), the second spectrum (shown in red) is recorded with two filters (A_{12}). This allows characterizing the transmission of the filter. **Right:** The measured transmission t_2 is shown in blue and the calculated transmission for unoxidized aluminum is shown in red. 74

Fig. 5.4 Beam profiles of the high harmonics transmitted through a thin-film foil of aluminum with imperfections. The profile in the left image shows some droplet-like structures, the filter in the right image shows a kink. These imperfections might have been created during manufacturing of the filters. 74

Fig. 5.5 Calculated group velocity versus photon energy for aluminum. Data of the refractive index has been taken from [79]. 75

Fig. 5.6 Calibrated high harmonic spectrum generated in argon. The total harmonic energy content per laser shot is 1.4 nJ. The conversion efficiency is $3.1 \cdot 10^{-6}$. 77

Fig. 5.7 Principle of the RABBITT technique. A sideband signal is generated in an electron time-of-flight spectrum. It is a result from interference between the absorption of a high harmonic photon with order q and an infrared photon (IR) and the absorption of a high harmonic photon with order q+2 and the emission of an infrared photon (IR). The sideband amplitude is a function of the delay between the XUV pump and IR probe pulse. 78

Fig. 5.8 Schematic of the experimental setup for the RABBITT measurement. A detailed description can be found in the text. 81

Fig. 5.9 Experimental realization of the optical setup for a RABBITT measurement in the high harmonic beamline. A detailed description can be found in the text. BS=beam splitter. 83

Fig. 5.10 Beam profile in the interaction region. Both IR beams are present and overlap in time and space. The two images show the situation for two different time delays: constructive (left) and destructive (right) interference. The situation shown here is good enough to allow for an attosecond time-resolved measurement. 84

Fig. 5.11 Harmonic spectrum (left axis) generated in argon with their corresponding relative delay, line as guide to the eye (right axis). 85

Fig. 5.12 RABBIT trace for a high harmonic generation in xenon. The photoelectrons have been generated in argon. The color scale was changed at a delay of ~ 4 fs for better visibility. 87

Fig. 5.13 (left) High harmonic spectrum (left axis) generated in xenon with the relative delays (right axis). The corrsponding RABBITT trace is shown in Fig. 5.12; **(right)** Reconstructed average pulse. 87

Fig. 6.1 Schematic of the experimental setup for the attosecond transient absorption measurement. 89

Fig. 6.2 Close up of harmonic 17. Signal without helium in the interaction (black), signal with helium where the IR pulse is 60 fs before the XUV (red) and 60 fs after the XUV (blue). The IR intensity is $1.3 \cdot 10^{13}$ W cm^{-2}. 91

Fig. 6.3 Close up of harmonic 15, same conditions as in Fig. 6.2. 92

Fig. 6.4 Relative absorption of harmonic 15, i.e. signal without helium divided by the signal with helium from Fig. 6.3 for the case where the XUV arrives 60 fs before the IR. 92

Fig. 6.5 Typical delay scan in logarithmic scale for better visibility. The infrared intensity was $1.3 \cdot 10^{13}$ W cm^{-2}. 93

Fig. 6.6 Spectrally integrated signal for harmonic 13 (left) and harmonic 15 (right) over the full spectrum of the harmonic. A clear structure on a fs timescale becomes visible. On top of it a faster modulation on an attosecond timescale is visible (see Sec. 6.5). The infrared intensity was $1.3 \cdot 10^{13}$ W cm^{-2}. ... 94

Fig. 6.7 Spectrally integrated signal of harmonic 17 which has a photon energy higher than the ionization potential of helium. In this case, the absorption is not significantly changed on a femtosecond timescale. The infrared intensity was $1.3 \cdot 10^{13}$ W cm^{-2}. 95

Fig. 6.8 Oscillations of the individual harmonic yield (orders 13, 15 and 17) around zero delay. The signal has been shifted vertically such that the harmonics appear underneath each other for better comparison of the relative phase differences. Data points are shown as dotted line, the solid line is a Fourier filtered signal, where oscillations faster than twice the fundamental laser frequency have been removed. The measurement has been taken with an aluminum filter thickness of 500 nm and an IR probe intensity of $1.3 \cdot 10^{13}$ W cm^{-2}. 96

Fig. 6.9 Oscillations of individual harmonics (orders 13, 15 and 17) around zero delay. The signal has been shifted such that the harmonics appear underneath each other for better comparison of the relative phase differences. Data points are shown as dotted line, the solid line is a Fourier filtered signal, where oscillations faster than twice the fundamental laser frequency have been removed. The measurement has been taken with an aluminum filter thickness of 500 nm and an IR probe intensity of $1.9 \cdot 10^{13}$ W cm^{-2}. 97

Fig. 6.10 A spectral integration of all the harmonics. The total yield shows an oscillation with half-cycle periodicity. The blue line corresponds to the experimental data, the red line is a smoothed version of the data (Fourier filtered). 99

Fig. 6.11 Relative absorption of harmonic 15, i.e. signal without helium divided by the signal with helium for the full delay scan. The absorption is strongest for zero time delay. At a delay of 20 fs (XUV before IR) a localized enhancement of the signal is measured. The enhancement factor reaches the value of two. The IR intensity was $1.3 \cdot 10^{13}$ W cm^{-2}. 100

Fig. 6.12 Relative absorption of harmonic 17, i.e. signal without helium divided by the signal with helium for the full delay scan. The average absorption is constant over the complete delay range. The IR intensity was $1.3 \cdot 10^{13}$ W cm^{-2}..**101**

Publications

Parts of this thesis have been published in the following journal papers and at conferences:

Journal Papers

1. M. Holler, F. Schapper, et al., "Attosecond Transient Absorption of High Order Harmonics in Helium," in preparation

2. F. Schapper, M. Holler, T. Auguste, et al., "Spatial fingerprint of quantum path interferences in high order harmonic generation," submitted

3. O. H. Heckl, C. R. E. Baer, C. Kränkel, S. V. Marchese, F. Schapper, M. Holler, T. Südmeyer, J. S. Robinson, J. W. G. Tisch, F. Couny, P. Light, F. Benabid, and U. Keller, "High Harmonic Generation in a Gas-Filled Hollow-Core Photonic Crystal Fiber," *Appl. Phys. B* 97, pp. 369–373, 2009

4. E. Cormier, A. Zair, M. Holler, F. Schapper, U. Keller, A. Wyatt, A. Monmayrant, I. Walmsley, T. Auguste, and P. Salières, "Direct observation of quantum-path interferences in high order harmonic generation," *European Physical Journal-Special Topics*, vol. 175, pp. 191-194, Aug 2009

5. T. Auguste, P. Salières, A. S. Wyatt, A. Monmayrant, I. A. Walmsley, E. Cormier, A. Zair, M. Holler, A. Guandalini, F. Schapper, J. Biegert, L. Gallmann, and U. Keller, "Theoretical and experimental analysis of quantum path interferences in high-order harmonic generation," *Phys. Rev. A*, vol. 80, 033817, 2009

6. E. Mansten, J. M. Dahlström, J. Mauritsson, T. Ruchon, A. L'Huillier, J. Tate, M. B. Gaarde, P. Eckle, A. Guandalini, M. Holler, F. Schapper, L. Gallmann, U. Keller, "Spectral signature of short attosecond pulse trains," *Phys. Rev. Lett., vol. 102, 083002 (2009)*

7. M. Holler, A. Zair, F. Schapper, T. Auguste, E. Cormier, A. Wyatt, A. Monmayrant, I. Walmsley, L. Gallmann, P. Salières, U. Keller, "Ionization effects on spectral signatures of quantum-path interference in high-harmonic generation," *Opt. Express, vol. 17, No. 7, pp. 5716-5722 (2009)*

8. A. Zair, M. Holler, A. Guandalini, F. Schapper, J. Biegert, L. Gallmann, U. Keller, A. S. Wyatt, A. Monmayrant, I. A. Walmsley, E. Cormier, T. Auguste, P. Salières, "Quantum Path Interferences in High-Order Harmonic Generation," *Phys. Rev. Lett., vol. 100, 143902 (2008)*

9. A. Zair, A. Guandalini, F. Schapper, M. Holler, J. Biegert, L. Gallmann, A. Couairon, M. Franco, A. Mysyrowicz, U. Keller, "Spatio-temporal characterization of few-cycle pulses obtained by filamentation," *Opt. Exp., vol. 15, pp 5394 (2007)*

Conference Papers

1. P. Rivière, U. Saalmann, M. Holler, F. Schapper, L. Gallmann, U. Keller and J.-M. Rost, "Transient absorption of high order harmonics from attosecond pulse trains", *International Workshop on ATOMIC PHYSICS, Dresden, Germany (2009)*

2. L. Gallmann, M. Holler, A. Zair, F. Schapper, T. Auguste, E. Cormier, A. Wyatt, A. Monmayrant, I. A. Walmsley, P. Salières, U. Keller, "Quantum path interferences in high-harmonic generation: spatial structure and spectral signature of ionization effects", invited talk, *Conference on Ultrafast and Nonlinear Optics (UFNO'09), Burgas, Bulgaria (2009)*

3. E. Mansten, J. M. Dahlström, J. Mauritsson, T. Ruchon, A. L'Huillier, J. Tate, M. B. Gaarde, P. Eckle, A. Guandalini, M. Holler, F. Schapper, L. Gallmann, U. Keller, "Spectral signature of short attosecond pulse trains," *Ultrafast Optics and High Field Short Wavelength 2009, Arcachon, France (2009)*

4. Mirko Holler, Florian Schapper, Johan Mauritsson, Kenneth J. Schafer, Lukas Gallmann and Ursula Keller, "Attosecond Transient Absorption of High Order Harmonics in Helium," *Second International Conference on Attosecond Physics July 2009*

5. Florian Schapper, Mirko Holler, Lukas Gallmann, Ursula Keller, "Chirp-controlled polarization gating for isolated attosecond pulse generation," *Second International Conference on Attosecond Physics July 2009*

6. O. H. Heckl, C. R. E. Baer, C. Kränkel, S. V. Marchese, F. Schapper, M. Holler, T. Südmeyer, U. Keller, J. S. Robinson, J. W. G. Tisch, F. Couny, P. Light, F. Benabid, P. St. J. Russell, "First demonstration of high harmonic generation in a hollow-core photonic crystal fiber," *Conference on Lasers and Electro-Optics – European Quantum Electronics Conference (CLEO Europe – EQEC) 2009*

7. O. H. Heckl, C. R. E. Baer, C. Kränkel, S. V. Marchese, F. Schapper, M. Holler, T. Südmeyer, U. Keller, J. S. Robinson, J. W. G. Tisch, F. Couny, P. Light, F. Benabid, P. St. J. Russell, "First demonstration of high harmonic generation in a hollow-core photonic crystal fiber," *Conference on Lasers and Electro-Optics (CLEO USA) 2009*

8. L. Gallmann, M. Holler, A. Zair, F. Schapper, T. Auguste, E. Cormier, A. Wyatt, A. Monmayrant, I. A. Walmsley, P. Salières, U. Keller, "Quantum path interferences in high harmonic generation: ionization effects and spatial structure," *Conference on Lasers and Electro-Optics (CLEO/IQEC USA) 2009*

9. F. Schapper, M. Holler, T. Auguste, A. Zair, M. Weger, P. Salières, L. Gallmann, U. Keller, "Spatially resolved quantum path interferences in high order harmonic generation in argon," *11th International Conference on Multiphoton Processes (ICOMP 2008)*

10. L. Gallmann, F. Schapper, A. Zaïr, A. Guandalini, M. Holler, A. Couairon, M. Franco, A. Mysyrowicz, and U. Keller, "Spatio-temporal effects in few-cycle filament compression," *Conference on filamentation, Paris, France September 2008*

11. A. Zair, M. Holler, F. Schapper, L. Gallmann, U. Keller, A. S. Wyatt, A. Monmayrant, I. A. Walmsley, E. Cormier, T. Auguste, J. P. Caumes, P. Salières, "Study of higher-order quantum-paths interference in the high harmonics generation process," *Ultrafast Phenomena 2008 (UP 2008) June 2008*

12. M. Holler, A. Zair, F. Schapper, T. Auguste, A. Guandalini, A. S. Wyatt, A. Monmayrant, L. Gallmann, I. A. Walmsley, E. Cormier, P. Salières, U. Keller, "Quantum-path interference in high harmonic generation," *SPS Annual Meeting 2008, Geneva, Switzerland, March 2008*

13. E. Mansten, J. M. Dahlström, P. Johnsson, M. Swoboda, T. Ruchon, A. L'Huillier, J. Mauritsson, J. Tate, M. Gaarde, P. Eckle, M. Holler, L. Gallmann, U. Keller, "Controlling attosecond pulse generation by using a two-color driving field," *402nd Wilhelm and Else Heraeus-Seminar on Novel Light Sources and Applications, Austria, 2008*

14. A. Zair, M. Holler, A. Guandalini, F. Schapper, J. Biegert, U. Keller, P. Salières, T. Auguste, E. Cormier, A. Wyatt, A. Monmayrant, I. Walmsley, "Direct Oberservation of quantum-path interenferences in high order harmonic generation," *1st International Conference on Ultra-intense Laser Interaction Sciences (ULIS 2007), Bordeaux, Oct 2007*

15. E. Cormier, M. Holler, A. Zair, A. Guandalini, F. Schapper, J. Biegert, T. Auguste, A. Wyatt, A. Monmayrant, I. Walmsley, P. Salières, U. Keller, "Direct observation of quantum-path interenferences in high order harmonic generation," *Ultrafast Optics 2007 and High Field Short Wavelength 2007 (UFO/HFSW 2007), Santa Fe, USA , Sep. 2007*

16. M. Holler, A. Zair, F. Schapper, T. Auguste, A. Guandalini, A. S. Wyatt, A. Monmayrant, L. Gallmann, I. A. Walmsley, E. Cormier, P. Salières, U. Keller, "A comparison of quantum-path interenferences in high order harmonic generation in Neon, Argon and Xenon," *Ultrafast Optics 2007 and High Field Short Wavelength 2007 (UFO/HFSW 2007), Santa Fe, USA, Sep. 2007*

17. F. Schapper, A. Courairon, A. Guandalini, A. Zair, M. Holler, M. Franco, A. Mysyrowicz, L. Gallmann, U. Keller, "Spatio-temporal structure and energy scaling of 4.9-fs pulses generated by filamentation," *Ultrafast Optics 2007 and High Field Short Wavelength 2007 (UFO/HFSW 2007), Santa Fe, USA, Sep. 2007*

18. A. Zair, M. Holler, A. Guandalini, F. Schapper, J. Biegert, U. Keller, A. Wyatt, A. Monmayrant, I. Walmsley, E. Cormier, T. Auguste, P. Salières, "Two-quantum-path interferences in high order harmonic generation," *XXV International Conference on Photonics, Electronic and Atomic Collisions (ICPEAC 2007), Freiburg, Germany, July 2007*

19. A. Zair, M. Holler, A. Guandalini, F. Schapper, J. Biegert, U. Keller, P. Salières, T. Auguste, E. Cormier, A. Wyatt, A. Monmayrant, I. Walmsley, "Two-quantum-path interferences in high order harmonic generation," *Conference on Lasers and Electro-Optics (CLEO Europe 2007), Munich, Germany, June 2007*

20. A. Zair, A. Guandalini, F. Schapper, M. Holler, J. Biegert, L. Gallmann, U. Keller, A. Couairon, M. Franco, A. Mysyrowicz, "Spatio-temporal characterization of sub-5-fs pulses obtained by filamentation," *Conference on Lasers and Electro-Optics (CLEO Europe 2007), Munich, Germany, June 2007*

21. A. Guandalini, A. Zair, F. Schapper, M. Holler, L. Gallmann, J. Biegert, U. Keller, "Spatio-temporal and interferometric characterization of sub-5-fs pulses obtained by filamentation," *Conference on Lasers and Electro-Optics (CLEO '07), Baltimore, USA, May 2007*

22. A. Zair, M. Holler, A. Guandalini, F. Schapper, J. Biegert, U. Keller, P. Salières, T. Auguste, E. Cormier, A. Wyatt, A. Monmayrant, I. Walmsley, "Two-quantum-path interferences in high order harmonic generation," *Conference on Lasers and Electro-Optics (CLEO '07), Baltimore, USA, May 2007*

23. L. Gallmann, T. Pfeifer, P. M. Nagel, M. J. Abel, D. M. Neumark, S. R. Leone, A. Zaïr, A. Guandalini, F. Schapper, M. Holler, U. Keller, "Few-cycle pulse generation through optical filamentation in rare gases and its spectral and spatio-temporal characterization," *Nonlinear Effects in Photonic Materials, Weierstrass Institute for Applied Analysis and Stochastics, Berlin, Germany, March 2007*

Abstract

The interaction of an ultrashort laser pulse with a gas at high intensity can lead to a coherent generation of high-order harmonics of the driving laser frequency. By this nonlinear interaction, the extreme ultraviolet (XUV) and soft-X-ray spectral regions become accessible. In time domain the generation of a train of pulses or isolated pulses with a duration in the attosecond (1 as=10^{-18} s) regime is possible. Thus, high-order harmonics can not only be used for spectroscopic applications, but also for time-resolved measurements. The attosecond pulse duration enables tracking electron dynamics inside atoms or molecules.

The high-order harmonic generation process is well described by a semi-classical model. An electron initially in the atomic ground state is freed into the continuum by tunnel ionization, accelerated, and ultimately driven back to the parent ion by the oscillating linearly polarized laser field. It may recombine to the ground state and release its kinetic energy plus the ionization potential of the atom by the emission of a photon. Electron trajectories with different excursion times can lead to the same photon energy. However, these photons carry a different harmonic phase.

In this thesis, the high-order harmonic generation process is investigated experimentally in temporal and spectral domain. For driving the generation process in noble gases, infrared (IR) laser pulses with a duration of 30 fs (1 fs=10^{-15} s) and a center wavelength of 800 nm (1 nm=10^{-9} m) are used.

In spectral domain, an interference structure between the two shortest electron trajectories, or quantum paths, is investigated. Proper experimental conditions permit a simultaneous spectrally resolved observation of the emission from both paths. A modulation of the harmonic yield with intensity of the driving laser is observed. These quantum path interferences (QPI) are measured for different generation media: neon, argon and xenon. This enables for the investigation of QPI with laser intensities below, around and above the intensity of barrier suppression, where the electron is set free without tunneling. In argon QPI is not only spectrally but also spatially resolved.

To enable attosecond time-resolved measurements, a versatile high harmonic beamline is developed and implemented. Typical attosecond experiments rely on a cross-correlation measurement of the combined XUV / IR fields with attosecond resolution. The beamline offers the required interferometric stability for these attosecond experiments. The interaction of the combined field with a target gas leads to the generation of photoelectrons. For their characterization a field-free electron time-of-flight spectrometer is developed and built. The new setup also features a beam profile and a flux measurement for the harmonics. The setup is equipped with an optical XUV spectrometer and allows for simultaneous detection of photons and photoelectrons.

Attosecond pulse trains (APT) generated in xenon and argon are temporally characterized by means of a RABBITT (reconstruction of attosecond beating by interference of two photon transitions) measurement. The IR field in the pump-probe measurement can lead to a two-photon process in the target medium. By energetically resolving the generated photoelectrons, the average attosecond structure of the APT can be reconstructed.

The APT generated in xenon is used in an all-optical XUV/IR pump-probe measurement. The gas density in the interaction region is increased to significantly absorb the harmonics. The absorption in helium is influenced by a strong IR field. The transmitted photon yield is spectrally resolved. It shows an envelope structure on a femtosecond timescale and a modulation on an attosecond timescale. This experiment can be regarded as the first all-optical XUV/IR pump-probe measurement with attosecond resolution. It is in good agreement with previous results based on the detection of the photo-ion yield. The experiment based on the spectrally resolved detection of photons gives further insight into the absorption dynamics close to the ionization threshold in helium. Due to the detection of photons, the experiment benefits from a fast acquisition time with excellent signal-to-noise ratio.

Kurzfassung (German)

Die Wechselwirkung eines ultrakurzen Laserpulses mit einem Gas kann bei hoher Intensität zu einer kohärenten Erzeugung von hohen Harmonischen der fundamentalen Laserfrequenz führen. Durch diesen nichtlinearen Prozess erhält man Zugang zum Spektralbereich des extremen Ultravioletts (XUV) und der weichen Röntgenstrahlen. Im Zeitbereich wird die Erzeugung von Pulszügen oder isolierten Pulsen mit einer Pulsdauer im Attosekundenbereich (1 as=10^{-18} s) möglich. Das ermöglicht die Verwendung der hohen Harmonischen nicht nur für Spektroskopie, sondern auch für zeitaufgelöste Messungen. Die Pulsdauer im Attosekundenbereich erlaubt es, Bewegungen von Elektronen in Atomen oder Molekülen zu verfolgen.

Die Erzeugung der hohen Harmonischen kann durch ein halbklassisches Modell beschrieben werden. Das Elektron ist zunächst im Grundzustand des Atoms. Es wird durch Tunnelionisation ins Kontinuum befreit, dort beschleunigt und schliesslich durch das oszillierende, linear polarisierte Laserfeld zu seinem Ion zurückgeführt. Dort kann es in den Grundzustand rekombinieren. Seine kinetische Energie plus das Ionisationspotential des Atoms werden dabei als Photon abgestrahlt. Hierbei gibt es mehrere Trajektorien des Elektrons mit unterschiedlichen Laufzeiten, die zur gleichen Photonenergie führen. Diese Photonen unterscheiden sich jedoch in ihrer Phase.

In dieser Doktorarbeit wird die Erzeugung von hohen Harmonischen im Zeit- und Spektralbereich experimentell untersucht. Zur Erzeugung der hohen Harmonischen in Edelgasen werden Laserpulse im Infraroten (IR) mit einer Dauer von 30 fs (1 fs=10^{-15} s) und einer Zentralwellenlänge von 800 nm (1 nm=10^{-9} m) verwendet.

So wird spektral eine Interferenzstruktur zwischen den zwei kürzesten Elektrontrajektorien, oder Quantenpfaden, untersucht. Geeignete experimentelle Bedingungen erlauben es, die Emission aus beiden Pfaden gleichzeitig zu betrachten. Hierbei wird eine Modulation des erzeugten harmonischen Flusses mit der Laserintensität beobachtet. Diese

Quantenpfadinterferenzen (QPI) werden in verschiedenen Erzeugungsmedien gemessen: in Neon, Argon und Xenon. Das ermöglicht eine Untersuchung mit Laserintensitäten unterhalb, um und über der Intensität der kompletten Potentialunterdrückung, wo das Elektron ohne zu tunneln befreit würde. Für Argon als Erzeugungsmedium wurden die QPI nicht nur spektral sondern auch räumlich aufgelöst.

Um Messungen mit Attosekunden Zeitauflösung zu ermöglichen, wird eine vielseitige Strahllinie entwickelt und realisiert. Attosekundenexperimente basieren typischerweise auf einer Kreuzkorrelationsmessung eines kombinierten XUV / IR Feldes mit Attosekunden Auflösung. Die neue Strahllinie hat die dazu nötige interferometrische Stabilität. Die Wechselwirkung des kombinierten Feldes mit einem Gas führt zur Erzeugung von Photoelektronen. Um diese zu charakterisieren wurde ein feldfreies Elektronen Flugzeitspektrometer entwickelt und gebaut. Die neue Anlage ermöglicht es ausserdem Strahlprofile und den Energiestrom der hohen Harmonischen zu bestimmen. Der Aufbau ist mit einem optischen XUV Spektrometer ausgestattet und erlaubt die gleichzeitige Messung von Photonen und Photoelektronen.

Attosekunden Pulszüge (APT), die in Xenon und Argon erzeugt wurden, werden mittels der RABBITT („reconstruction of attosecond beating by interference of two photon transitions", sprich „Rekonstruktion eines Attosekunden Signals der Interferenz eines Zweiphotonenübergangs") Technik zeitlich charakterisiert. Das IR Feld in der Anregung-Probe Messung kann zu einem Zweiphotonenprozess im Wechselwirkungsmedium führen. Die mittlere zeitliche Attosekunden Struktur in dem APT kann durch energieaufgelöste Messung der erzeugten Photoelektronen rekonstruiert werden.

Der APT, der in Xenon erzeugt wurde, wird in einer komplett optischen XUV/IR Anregung-Probe Messung verwendet. Dabei wird die Dichte des Gases in der Wechselwirkung soweit erhöht, dass die Harmonischen zu einem wesentlichen Teil absorbiert werden. Die Absorption in Helium wird hierbei von dem starken IR Feld beeinflusst. Der transmittierte Photonenfluss wird spektral aufgelöst. Er zeigt eine Femtosekundenstruktur auf der Einhüllenden sowie eine Modulation auf einer Attosekunden Zeitskala. Dieses Experiment kann als die erste komplett optische XUV/IR Anregung-Probe Messung mit Attosekunden Zeitauflösung betrachtet werden. Es stimmt gut mit älteren experimentellen Ergebnissen, die auf der Messung der erzeugten Photoionen basieren, überein. Die spektral aufgelöste Messung der Photonen gibt allerdings weiteren Einblick in die Dynamik der Absorption nahe am Ionisationspotential von Helium. Aufgrund der direkten Messung von Photonen profitiert das Experiment von einer schnellen Datenerfassung mit einem sehr guten Signal-zu-Rausch-Verhältnis.

Chapter 1

Introduction

Since the first demonstration of a laser by T. H. Maimann in 1960 [1], laser technology has undergone a rich development and many applications for laser sources have been found. At the beginning, these laser sources produced either continuous light waves or pulses with a duration in the microsecond regime. The invention of Q-switching, and later mode-locking techniques allowed creating shorter and shorter pulses, which gave the possibility to investigate faster and faster processes in nature. Nowadays femtosecond pulses are routinely available from laser oscillators. These ultrashort pulses allowed chemists to temporally resolve bond breaking or bond formation in chemical reactions [2, 3], which occur on a femtosecond timescale. In the spectral domain, the frequency comb of a modelocked laser has been used for high-resolution spectroscopy [4, 5].

The resolution of processes on a faster timescale, like the electronic motion in an atom or molecule, requires sub-fs pulses. The pulse duration achievable by lasers in the visible or IR spectral range is limited by their spectral bandwidth. The generation of sub-femtosecond pulses entails a spectral broadening accompanied by a reduction of the central wavelength into the XUV or soft-X-ray spectral region.

One possible way to produce shorter pulses is the generation of high-order harmonics of a laser field in a solid or gaseous target [6, 7]. HHG is a coherent up-conversion process [8], offering the potential to create pulses or trains of pulses [9] with a duration in the 100 as regime and less [10]. HHG is a non-perturbative nonlinear process and can be visualized by a semi-classical model [11, 12]: the atom irradiated with the laser will be tunnel ionized by the strong electric field of the laser. The freed electron propagates in the laser field and thereby gains kinetic energy. Upon return it may recombine with its parent ion in a radiative process. The energy of the emitted photon can reach the keV regime, which has been used for X-ray spectroscopy [13]. The photon contains the energy of the electron plus the ionization potential of the atom. Short thereafter, the model has been confirmed by a fully quantum mechanical formulation [14, 15].

The model of HHG predicts that different electron trajectories of the free electron can generate the same photon energy. However, these photons carry a different harmonic phase [16]. Several types of electron trajectories have been investigated separately [17, 18]. They can be selected by exploiting phase-matching techniques [19]. When several trajectories contribute simultaneously to the macroscopic signal, an interference structure in the high harmonic signal can be expected, which reveals itself as a modulation of the harmonic yield with laser intensity, the QPI (quantum path interferences).

For efficient tunneling ionization, the electric field of the driving laser has to be comparable to inner-atomic Coulomb field strengths [20]. Furthermore, the maximum energy the electron can reach is proportional to the laser intensity [12]. Therefore, the generation of high harmonics requires laser intensities in the range of 10^{14} W cm^{-2}. These became accessible with the invention of the chirped pulse amplification (CPA) technique [21]. It allows producing laser pulses with pulse energies in the millijoule regime and above whereas pulses obtained from an oscillator directly typically have an energy in the nanojoule regime.

The high harmonic generation process is very inefficient due to the small probability of recombination. The conversion efficiencies are in the range of 10^{-5} and less. Nonetheless, it was possible to measure a second-order autocorrelation of high harmonics with attosecond resolution in helium [22]. In this case, the harmonics had photon energies below the ionization potential of helium. For harmonics with higher energy, the medium is ionized by a single photon process which makes it difficult to further extend this autocorrelation technique. At high energy, the relevant two-photon cross section becomes too small [23]. However, it is possible to perform time-resolved measurements in XUV/IR pump-probe schemes, where a replica of the driving laser field together with the harmonic radiation is used. This is possible because of the inherent synchronization of the driving field with the harmonic emission [24]. The intensity of the replica of the driving pulse can be very high, eliminating the need of a high intensity XUV pulse. This enabled the temporal characterization of APTs [25] and of isolated attosecond pulses [26, 27].

These pulses have been used to access sub-fs electron dynamics in atoms and molecules. With suitable photon energies it is possible to excite core-levels, which triggers numerous intra-atomic relaxation processes. The transient inner-shell hole can be filled with an electron from an outer shell. Thereby the binding energy can be released by a photon or second electron (Auger electron). This way it was possible to trace dynamic information on the lifetime of M-shell vacancies of krypton directly in time domain with attosecond resolution [28]. In another type of experiment, the spectral response of HHG has been used directly to obtain information on the structure and dynamics of the bound electronic wave functions in the generating medium. These interfere with the returning wave packet of the free electron and exhibit a signature in the high harmonic spectrum. The dependence of the harmonic yield on the angle between the molecular axis and the polarization of the driving laser field is seen to contain the fingerprint of the highest occupied molecular orbitals. It enabled the tomographic reconstruction of the highest occupied molecular orbital of N_2 [29]. Later, the technique was extended beyond the simple diatomic molecules to larger polyatomic molecules [30].

In this thesis, a detailed description of the HHG process will be given in chapter 2. The theoretical model of HHG will be presented and its implications on the intrinsic properties of the harmonic radiation will be discussed on both, the microscopic and the macroscopic level. In addition, the CPA laser system, which was used in this thesis, will be described.

Experimental conditions are found that permit for the observation of QPI. In chapter 3, it will be demonstrated that a change in laser intensity can be used to control QPI. A change in the interference structure corresponds to a change in the relative timing of the electron quantum paths, which occurs on an attosecond timescale. QPI is measured in neon, argon and xenon which allows for the investigation of QPI with laser intensities below, around and above the intensity of barrier-suppression.

QPI permits obtaining information on the HHG process itself. It is a purely spectral measurement. The attosecond emission is not used directly to achieve the high temporal resolution. To enable attosecond pump-probe measurements at ETH, a new, versatile attosecond setup has been designed and implemented. It offers the required mechanical stability and can be easily adopted for different types of experiments. It is described in detail in chapter 4. The setup has been used in several time-resolved measurements. In typical attosecond experiments, photoelectrons generated in a target medium are energetically resolved. For this purpose, a field-free electron time-of-flight spectrometer has been designed and built. In the new setup it is possible to operate the time-of-flight spectrometer and a XUV photon spectrometer simultaneously.

The setup is shown to work for attosecond experiments in chapter 5. APTs generated in argon and xenon have been characterized by means of a RABBITT measurement [25]. An attosecond beating of a two-photon process is measured in a XUV/IR pump-probe measurement which allows for the reconstruction of the average pulse duration of the attosecond bursts in the APT. In addition, beam profiles of harmonic radiation as well as the harmonic flux have been measured in the new harmonic beamline.

In chapter 6, a new type of attosecond experiment is presented: The attosecond transient absorption of an APT generated in xenon. Helium is used as target medium in a XUV/IR pump-probe measurement. The typical detection of photoelectrons or –ions has been replaced by a direct detection of photons. The density of the interaction target was increased to significantly absorb the APT. The transmitted XUV has been detected in the XUV spectrometer. Delay dependent structures on the femtosecond timescale as well as on the attosecond timescale are found. This experiment is the first all-optical XUV/IR pump-probe measurement with attosecond resolution. Our results are in good agreement with a similar experiment, where the yield of the generated photo-ions has been measured [31]. In our measurement we gain further insight into the absorption dynamics close to the ionization threshold in helium. Due to the photon detection the experiment benefits from a good signal-to-noise ratio and from a fast acquisition time.

Chapter 2

High Harmonic Generation

The generation of high-order harmonics of a low-frequency laser field has been widely studied in the recent years and creates a light source in the XUV/soft-X-ray spectral region that is spatially and temporally coherent. The broad bandwidth of high harmonic radiation enabled the generation of pulses with duration in the attosecond regime, which allows for experiments with high temporal resolution. In this chapter the principles of and requirements for high harmonic generation will be presented.

2.1 High harmonic generation

High harmonic generation (HHG) [6] is a process in which an intense laser pulse (laser intensities typically range from 10^{14} to 10^{16} W cm^{-2}) interacts with an atomic or molecular gas and odd harmonics $N\omega$ of the laser frequency ω up to some cutoff order N_{max} are emitted in the forward direction. Thereby N_{max} can reach numbers of several hundreds. If the photon energy of the driving laser is much smaller than the ionization potential of the atomic medium, the harmonic spectrum has a very characteristic and universal shape: it falls off exponentially for the first few harmonics, then exhibits a plateau where all the harmonics have nearly the same strength, and ends with a sharp exponential cutoff.

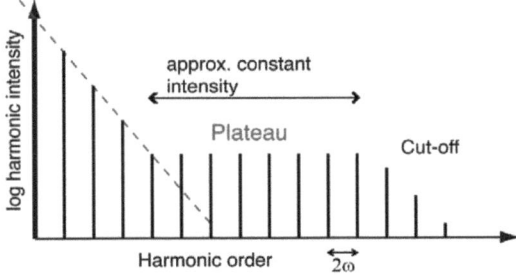

Fig. 2.1 Typical high harmonic spectrum generated by a multi-cycle driving field showing the plateau with nearly constant intensity, followed by the cut-off.

This means that high harmonic generation is not a perturbative nonlinear (i.e., is not a $X^{(N)}$) process for which each successively higher order would be expected to be emitted with a smaller efficiency. Indeed, for laser intensities of 10^{14} W cm^{-2} the corresponding electric field has an amplitude of $\sim 10^8$ V cm^{-1}, which cannot be treated as a small perturbation since this amplitude reaches values close to typical inner-atomic Coulomb-field strengths (first Bohr orbit $E_C = 5.1 \times 10^9$ Vcm^{-1}).

The Keldysh parameter [32]

$$\gamma = \frac{\hbar \omega}{e a_{\text{Bohr}} E}$$

can be used to estimate if a process can be treated in a perturbative way, which is the case for $\gamma \gg 1$ (multiphoton processes dominating), while for $\gamma \ll 1$ the process will be non-perturbative (tunneling regime). e and m are the charge and rest mass of the electron, $a_{\text{Bohr}} = \hbar / \sqrt{2mI_p}$ is the Bohr radius, I_p is the ionization potential of the atom and E is the amplitude of the external electric field.

For understanding the physics of the harmonic emission one has to consider two aspects:

- The microscopic emission is caused by the nonlinear response of the atoms subjected to the intense laser field. This will be discussed in Sec. 2.1.1.

- The macroscopic harmonic field (which can finally be measured) is the coherent superposition of the fields emitted by all the atoms of the generating medium and will be discussed in Sec. 2.1.4.

2.1.1 Single atom response

The key process triggering strong-field phenomena is ionization. Depending on the intensity of the driving electrical field, there are different regimes of how an atom can be ionized. Exposing an atom to an intense laser field will result in a modified potential, which is a combination of the unperturbed Coulomb potential and the time-dependent field of the optical pulse. This is schematically shown in Fig. 2.2 for different field strengths of the driving electric field.

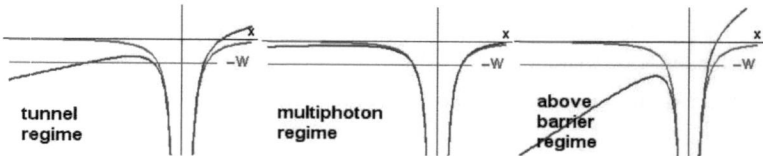

Fig. 2.2 Regimes of atomic ionization by low energetic photons ($E_{photon} < I_p$), where $W = I_p$ is the ionization potential of the atom.

At moderate intensities the resulting potential is close to the unperturbed Coulomb potential and an electron can be liberated only upon simultaneous absorption of N photons, resulting in multiphoton ionization ($\gamma \gg 1$). At sufficiently high field strengths the Coulomb barrier becomes narrow ($\gamma < 1$), allowing tunnel ionization to take over, resulting in a tunneling current that follows adiabatically the variation of the resultant potential. If the field strength becomes even higher, the electric field amplitude will suppress the Coulomb barrier below the energy level of the ground state, opening the way to above-barrier ionization. For laser intensities higher than the barrier suppression intensity (BSI), the electron can escape even classically (without tunneling) from the atom. This intensity can be calculated in first approximation by a simple model which consists of the Coulomb potential with the additional static electric field. This results in $I_{BSI}[\text{W/cm}^2] = 3.8 \cdot 10^9 \cdot I_p^4[\text{eV}]/Z^2$ [33], with I_p being the ionization potential of the atom, and Z the ionic charge. Finally, at intensities higher than 10^{17} W cm^{-2} we enter the relativistic regime [34], where the magnetic field of the driving laser field becomes relevant and cannot be neglected anymore.

In the intensity range of interest, multiphoton ionization, and/or tunneling ionization play the predominant role. In the case where tunneling ionization is the leading ionization-process, many of the features of high-harmonic generation can be understood in terms of a semi-classical model developed by Kulander [11] and Corkum [12]. They explain the emission of high-harmonic radiation in terms of three discrete steps illustrated in Fig. 2.3:

1. The electron tunnels through the barrier formed by the atomic Coulomb potential and the laser field. It is set free with zero initial velocity.

2. The quasi-free electron subsequently acquires kinetic energy from the laser field, and within an optical cycle, it returns (accelerated by the linearly polarized field) to the parent ion.

3. The electron recombines with some (small) probability to the ground state, emitting a photon with the energy of the ionization potential of the atom plus the kinetic energy gained in the laser field.

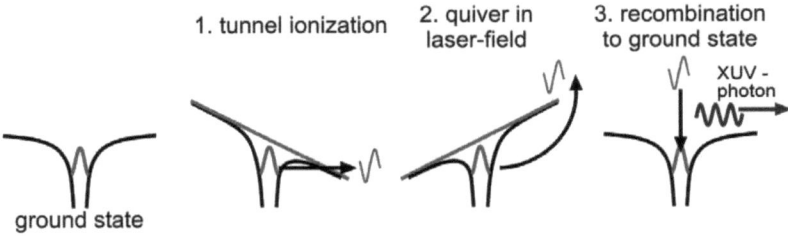

Fig. 2.3 Illustration of the three-step-model [12]. See text for details.

One important breakthrough of this model is the understanding of the cutoff-behavior in high-harmonic spectra. In 1992 a *cutoff law* for the extension of the harmonic plateau as a function of the parameters of the interaction was empirically discovered by Krause and co-workers by extensive numerical calculations [35]. Short time later this law was explained by Kulander [11] and Corkum [12] in terms of the semi-classical model. One can find that the cutoff in the harmonic spectrum occurs for harmonics of orders higher than $N_{max}\hbar\omega = I_p + 3.17 U_p$, where $U_p = e^2 E^2 / 4m\omega^2$ is the ponderomotive energy (i.e., the mean kinetic energy) of the electron quivering in the laser field. As a result, the extension of the plateau increases linearly with the laser peak intensity.

It is important to realize that in this model the process of high-harmonic generation starts with atoms in the ground state. Once the electron is set free, it will respond linearly to the laser field. This means that we do not only need high intensities for generating high harmonics, but also laser pulses of short duration: In the case of long, multi-cycle, high-intensity (such that tunneling ionization could come into play) laser pulses, the field strength will grow slowly and many atoms are ionized due to multiphoton ionization over many optical cycles. As a result, the ground state is depleted long before the peak of the laser pulse. However, for few-cycle pulses this saturation of ionization is shifted to considerably higher intensities. Only in the 10-fs regime does multiphoton preionization become negligible and optical-field ionization fully takes over. If the pulses are too long, N_{max} is not determined by the peak intensity, but by the laser intensity at which the ground state is depleted on the leading edge of the pulse. For shorter pulses, ground-state depletion is shifted to higher intensities [36].

2.1.2 Electron trajectories and intrinsic harmonic phase

Radiation generated in the high harmonic generation process carries an intrinsic phase, which can be explained by the semi-classical model. We calculate trajectories that an electron takes after tunneling using classical mechanics. The trajectories depend on the phase and on the polarization of the driving electric field $E(t) = E_0 \cos(\omega t + \phi)$. The electron is set free at time $t = 0$. Since a typical atomic scale is much shorter than the wavelength of the driving laser, one can estimate the electric field to be constant over the spread of the free electron. Furthermore, we will only treat the case of linearly polarized light here. In Fig. 2.4 plots of the electric field (blue), and location of the free electron for different electron release phases of the electric field versus time are shown. For a phase $\phi \in [0, \pi/2]$, there will be at least one return of the electron to the nucleus (green) where a recombination to the ground state can occur. For phases $\phi \in (\pi/2, \pi)$ the electron will not return to the nucleus (red), so no harmonic emission can be expected from these trajectories.

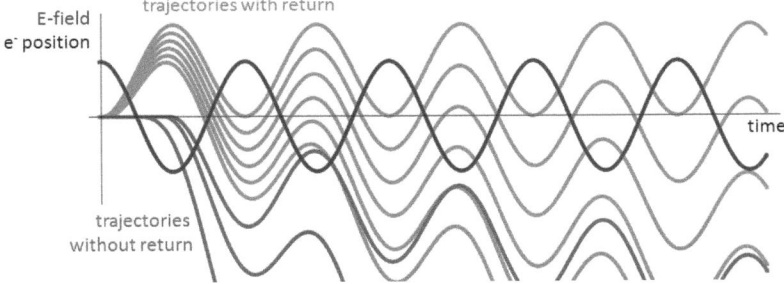

Fig. 2.4 Electric field (blue) and position of the electron for different electron-release-phases versus time (arb. u.). Depending on the release phase there are trajectories that return to the ion (green), and trajectories that do not return (red).

In the case that a trajectory experiences more than one possible return to the nucleus, the probability for recombination to the ground state is highest for the first return due to quantum diffusion of the free electron wave packet in the continuum. This, of course, is not explained by the semi-classical model. From this point on, we will consider the first recollision only.

The energy of the emitted photon depends on the energy the electron has at its return to the nucleus. This is reflected in Fig. 2.5 (for returning phases only). The electron trajectories are shown for different release phases like in Fig. 2.4. The electric field is shown in blue and the electron trajectories are color coded corresponding to the energy the electron will have upon its first return. One can find a single trajectory with highest return energy of 3.17 U_p, which is the cutoff trajectory. It is reached for a release phase of $\phi = 0.313$ rad $\approx 17°$. For lower energies, two trajectories exist. Though their energy is identical they differ in excursion time τ, describing the time the electron spent in the continuum.

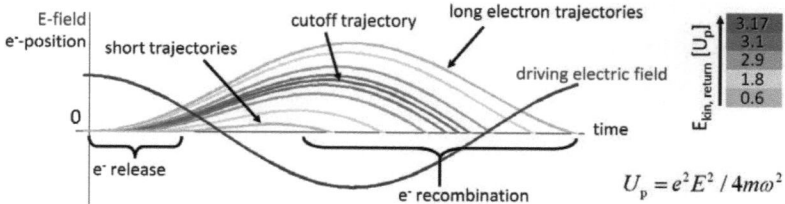

Fig. 2.5 Classical electron trajectories and their energy upon return to the nucleus. One cutoff trajectory can be identified. For lower energies two trajectories exist, one with a long and one with a short excursion time.

We can further investigate this fact by calculating the energy-spectrum of the returning electron versus phase of electron-release as shown in Fig. 2.6. The red line corresponds to the electron-kinetic energy, the black line shows the corresponding excursion time. One can identify two different pairs of trajectories, one with a short excursion time τ_{short} and one with a long excursion time τ_{long}. These pairs are also indicated in Fig. 2.5. Both fulfill $0 < \tau_{short} < \tau_{long} < T$, with T being the laser period.

The different excursion times lead to a different phase of the generated harmonic photons. In a first approximation, the harmonic phase is proportional to the product of minus the mean kinetic energy $-U_p$ and the excursion time τ [37]. Thus the intrinsic phase is different for the 'long' or 'short' trajectory and will become important for the macroscopic emission discussed in Sec. 2.1.4 and for the generation of attosecond pulses discussed in Sec. 2.1.5 and Sec. 5.4.

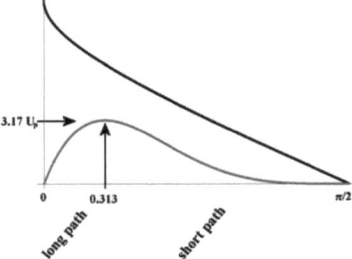

Fig. 2.6 Energy-spectrum of the returning electron versus phase of electron-release (red). The black line shows the time the electron spent in continuum (arb. u.).

The laser-induced phase $-U_p\tau$ will also lead to a phase modulation (chirp) of the harmonic emission due to the time dependence of $U_p = U_p(t)$. The harmonic spectrum exhibits a signature of the τ_{short} and τ_{long} trajectories. This aspect will be discussed in more detail in Chapter 3.

The last item addresses the next step to follow in the classical model to find the shape of a high harmonic spectrum (Fig. 2.1): Now that the energy distribution and the excursion times of the electrons are found, one has to weight these with the probability of the electron tunneling out (phase dependent, e.g. ADK tunneling rates [32]) and the probability of the electron recombining with the core to the ground state. However this will not be worked out here.

So far the theory was treating linearly polarized light only. If the light exhibits an elliptic component the process will become more and more inefficient. In the classical model there will be no return of the electron back to the nucleus for most starting phases. However in reality a small ellipticity is acceptable, but the amplitude of the generated harmonic signal will decrease by some decades as the ellipticity grows. This has been studied in detail experimentally and theoretically in [38, 39].

2.1.3 Quantum mechanical formulation

In 1994 a full quantum theory of high-harmonic generation by low-frequency laser fields [14] (valid in the tunneling limit) was presented. This theory recovers the classical interpretations explained above and also explains why the single-atom harmonic spectra fall off at an energy described by the cutoff-law. It is based on solving the time-dependent Schrödinger equation and will be presented here in a short summary.

The atom is considered in a single-electron approximation (atomic potential $V(x)$) under the influence of the laser field with linear polarization in x-direction $E(t) = (E\cos(\omega t), 0, 0)$. The Schrödinger equation writes

$$i\hbar \frac{\partial}{\partial t}|\psi(\vec{x},t)\rangle = \left[-\frac{\hbar^2 \nabla^2}{2m} + V(\vec{x}) - E\cos(\omega t)ex\right]|\psi(\vec{x},t)\rangle.$$

Solving this problem is very demanding and can be done numerically. To calculate a high harmonic spectrum, the x-component of the time-dependent dipole moment must be calculated

$$ex(t) = \langle\psi(\vec{x},t)|ex|\psi(\vec{x},t)\rangle.$$

The harmonic components of the spectrum are given by the Fourier transform of the dipole acceleration [40].

Assumptions can be made that simplify the problem of calculating $x(t)$. These will also allow gaining additional insight into the physics of the high harmonic generation process:

a. The system can be treated without any intermediate states, such that it only contains the ground state $|0\rangle$ and the continuum states.

b. The depletion of the ground state can be neglected, as long as we ionize the medium weakly (i.e. for laser intensities smaller than BSI).

c. The amplitude of the laser field dominates over the Coulomb field of the ion for the ionized electron. The electron can be treated as a free particle which is moving in the electric field of the laser without being influenced by the atomic potential $V(\vec{x})$. This simplification is called strong field approximation (SFA). This approximation can be justified because the electron will tunnel out when the laser field peaks, it will be accelerated immediately by a strong field and leave the region of the ion. When it comes back to the nucleus it will have a large kinetic energy.

Assumption a. and c. allow writing the time-dependent wave function $|\psi(t)\rangle$ as an expansion into a ground state component $|0\rangle$ with amplitude $a(t)$ and continuum components $|\vec{q}\rangle$ with amplitude $b(\vec{q},t)$:

$$|\psi(t)\rangle = e^{iI_p t/\hbar}\left(a(t)|0\rangle + \int d^3q\, b(\vec{q},t)|\vec{q}\rangle\right).$$

The continuum states correspond to outgoing electrons with kinetic momentum $\vec{q}$ and are eigenstates of the free Hamiltonian where $E(t)=0$.

By neglecting the ground state depletion (assumption b.) we can set $a(t)=1$.

The Hamiltonian can be divided into two parts, one part describes the effects of the motion of the electron in the external field, the other part describes rescattering effects, which can be neglected for HHG. The Schrödinger equation for $b(\vec{q},t)$ reads then

$$\hbar \dot{b}(\vec{q},t) = -i\left(\frac{\vec{q}^2}{2m}+I_p\right)b(\vec{q},t) + iE(t)\cos(\omega t)d_x^*(\vec{q}) - \hbar e E(t)\cos(\omega t)\frac{\partial b(\vec{q},t)}{\partial q_x}$$

and its solution is of the form

$$b_0(\vec{p},t) = i\int_0^t dt' E(t')\cos(\omega t')d_x(\vec{p}-e\vec{A}(t')/c)$$

$$\cdot \exp\left(-i\int_{t'}^t dt''\left[\left(\vec{p}-e\vec{A}(t'')/c\right)^2/2m + I_p\right]/\hbar\right). \qquad (1.1)$$

Here, $d_x(\vec{q})$ denotes the component of the atomic dipole matrix element for the bound-free transition $(\vec{d}(\vec{q}) = \langle \vec{q}|e\vec{x}|0\rangle)$ that is parallel to the polarization axis, which now contains all information about the atom. $\vec{A}(t)$ is the vector potential and $\vec{p} = \vec{q} + e\vec{A}(t)/c$ is the canonical momentum.

At that point we can explain the physics that is described by equation (1.1): $b_0(\vec{q},t)$ describes a sum of probability amplitudes that the electron is born in the continuum at time t' with canonical momentum $\vec{p}$. The electrons propagate until time t. During propagation they acquire a phase factor of $\exp(-iS(\vec{p},t,t')/\hbar)$, with $S(\vec{p},t,t')$ being the quasi-classical action. This action describes the motion of an electron that is moving in the laser field as a free particle with constant canonical momentum $\vec{p}$ and is given by

$$S(\vec{p},t,t') = \int_{t'}^{t} dt'' \left[\left(\vec{p} - e\vec{A}(t'')/c \right)^2 / 2m + I_p \right].$$

The recombination (free-bound transition) is given by $d_x^*(q)$. For calculating the high harmonic spectrum, we need to calculate the time-dependent dipole moment, which is now given by

$$e\vec{x}(t) = \int d^3 \vec{q}\, d_x^*(\vec{q}) b_0(\vec{q},t) + \text{c.c.}.$$

Later it was realized that this equation can be further analyzed by using Feynman's formulation of quantum mechanics [15, 41] since the equations contain a transition element from the ground state to the continuum, then propagate the electron in the continuum (with a propagator of i times the quasi-classical action) and finally there is the transition from the continuum back to the ground state. By using a saddle point method to calculate the integrals of the propagation one can find that only few quantum orbits (space-time trajectories) have a significant contribution to the final result. The expressions can be transformed into sums. Their contributions correspond to quasi-classical electron trajectories, characterized by the time of birth, their canonical momentum and the time of recombination. Their real part corresponds to the trajectories of the semi-classical model. One can show that the n-th Fourier component of the dipole moment can be expressed as a sum over relevant electron trajectories, with a_k being their amplitude and S the corresponding action

$$ex_n = \sum_k a_k e^{-iS(p_k,t_k,t_k')/\hbar}.$$

By this the semi-classical model of high harmonic generation (Sec. 2.1.1) and its results (Sec. 2.1.2) are confirmed to be an adequate description of the generation process and the high harmonic emission.

2.1.4 Macroscopic response

Typical harmonic spectra will show a different behavior than expected from the description given in the previous sections. For example, they might show a cutoff at much lower energy than predicted by the cutoff-law or the harmonics in the plateau region do not have equal intensity as expected from the single-atom response. Like in perturbative nonlinear optics, the macroscopic harmonic field in the tunnel regime is the coherent superposition of the fields emitted by all the atoms of the generating medium.

The macroscopic harmonic emission is optimized when phase matching is achieved. For each harmonic order q with energy $\omega_q = q\omega$ the principle of momentum conservation (which is in the easiest case: $\vec{k}_q = q\vec{k}$) has to be fulfilled, $\vec{k}_q$ being the wave vector of the harmonic photon and $\vec{k}$ the wave vector of the fundamental. This means that the sum of the participating wave vectors has to add to zero for perfect phase matching. Otherwise there is a mismatch in k of $\Delta k = |\vec{k}_q - q\vec{k}|$. In this case, the coherence length $L_{coh} = \pi / \Delta k$ will be finite and the harmonic amplitude will oscillate as a function of interaction length with period $2L_{coh}$. In the case of perturbative nonlinear optics (e.g. second harmonic generation) effective phase matching methods can be achieved, for example by matching the refractive index of the fundamental to the harmonic in a nonlinear birefringent crystal. No effective technique for achieving phase matching is available for high harmonic generation. The usable length of the generating media is usually limited by reabsorption (the harmonic photons ionize the generation medium in a single photon process).

Additional phase effects occur in high harmonic generation:
- We have tightly focused light pulses. A Gaussian beam acquires an additional (geometric) phase which differs from that of a plane wave. It is given by $\varphi(z) = -\arctan(z/z_R)$, with z being the propagation direction (focal position at $z=0$) and z_R the Rayleigh length (i.e. the distance from $z=0$ to where the focal diameter is increased by $\sqrt{2}$). This phase difference is called Gouy phase shift [42].

- We have an intrinsic dipole-phase which is (as mentioned above) in first approximation $-U_p \tau$. For each harmonic it can take two values, τ_{short} and τ_{long}. Furthermore (since $U_p \propto E^2$) it has a spatial dependence because of the spatial laser intensity profile $I(r,z)$ with z being the propagation direction, and $r \perp z$. It also has a temporal dependence due to the time structure of the laser pulse $I(t)$.

- In the generation process one generates free electrons which induces a variation of the refractive index [43] in the medium proportional to the local density of free electrons: $\Delta n(r,z,t) \propto -N_e(r,z,t)$. This results in a strong normal dispersion because of the time dependence of the laser intensity. As a result, the spectrum of the laser can be shifted to the blue (self-phase modulation) [44]. Spatially, the density gradient of free electrons due to the spatial intensity profile, causes a defocusing of the laser beam, decreasing the peak intensity [45].

Phase matching may be achieved for some positions in the focal region [46] where the individual contributions compensate each other. This can be reached either on axis ($r \approx 0$) or off axis ($r \neq 0$). In the longitudinal direction the position-dependent dipole phase is a symmetric term (relative to the focus position). It competes with the antisymmetric geometric phase (Gouy phase). The 'short' trajectory (τ_{short}) will be selected for a focusing before the jet and phase matching will be achieved on-axis. The 'long' trajectory (τ_{long}), associated with a dipole phase rapidly varying with intensity, is favored for a focusing after the gas jet and will lead to good phase matching off-axis. In this case, however, the short trajectory might also be partly phase-matched (on-axis). A detailed discussion about phase matching in terms of wave-vectors has been performed by Philippe Balcou and co-workers in [19].

The strong dependence of phase matching on the position of the generating medium with respect to the laser focus will have direct consequences for the:

- far field high-harmonic spatial intensity profile: Focusing before the gas jet will result in a narrow, well shaped profile with intensity maximum at $r = 0$, while focusing after the jet can lead to a diffuse intensity profile.

- spectral profile of the harmonics: Focusing before the gas jet will prefer the short trajectory in the macroscopic harmonic emission (of course both trajectories are present on atomic scale), while focusing after the jet will prefer the long trajectory.

A resulting harmonic emission consists of many contributions, each with different time-frequency characteristics. Thus the macroscopic behavior is not easy to predict and understand. It depends on many factors like focusing, gas density, interaction length, and temporal characteristics of the driving-pulse.

2.1.5 Temporal structure of high harmonics

So far we discussed high harmonic generation in spectral domain. In this chapter we will discuss the temporal structure of the high harmonics, especially the emission of light bursts with attosecond duration and the formation of attosecond pulse trains.

The shortest pulse duration T_{min} that can be achieved with a laser source with center wavelength λ is limited by the oscillation period of the light. In the case of a IR pulse centered at $\lambda = 800\,nm$, the duration of a single optical cycle is $T_0 = 2\pi/\omega = 2.7\,fs$. The up-conversion of a laser field using high harmonic generation shortens the wavelength. The effective spectral bandwidth increases to $\approx N\omega$ (N being the cut-off harmonic order). Since high harmonic generation is a coherent process, pulses with a duration of $\tau \approx 2\pi/N\omega$ can be generated, which will be in the attosecond time regime [47].

In section 2.1.1 we discussed the semi-classical model. In Fig. 2.6 the electron return energy for the two shortest electron trajectories is shown. From this model one expects a continuous harmonic spectrum generated by each half cycle of the driving electric field. It is possible to restrict the emission to a single trajectory by selecting proper phase-matching conditions or by applying a spatial filter (see Sec. 2.1.4). In that case, the emission generated by each optical half-cycle will have a broad continuous spectrum with a defined phase. In time domain a single pulse with a duration of few hundred attoseconds can be formed.

In the case of a multi-cycle driving field, high harmonic bursts will be emitted each half-cycle of the driving field. The radiation will consist of a sequence of pulses separated by half the optical period.

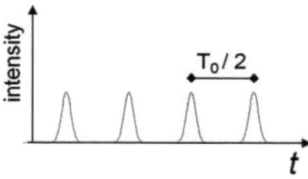

Fig. 2.7 A multi-cycle driving pulse will generate a train of attosecond pulses in time domain. The separation of the pulses is one half-cycle of the driving field.

The Fourier transform of this (coherent) pulse train is a series of components separated by twice the laser frequency. Thus, the spectrum of such an attosecond pulse train (APT) will consist of odd harmonics only as indicated in Fig. 2.1 [9, 47]. A more detailed description of attosecond pulse trains and their properties will be given in section 5.4.

2.2 Laser systems

In the previous section we have seen that for the generation of high-order harmonics short laser pulses with high intensities are required. Assuming a laser pulse duration of 30 fs and a laser focus of 60 µm diameter the pulse energy required to achieve an intensity of 10^{14} W cm^{-2} is 80 µJ. In an experiment much higher pulse energies might be necessary due to the losses in the experimental setup. Today, Ti:Sapphire laser systems are used in most experiments working on high harmonics. They are a proven technology for femtosecond pulse generation at a wavelength of around 800 nm. In section 2.2.1 the Ti:Sapphire laser will be described in general. In section 2.2.2 a more detailed description of the specific laser system used in the experiments of this thesis will be given.

2.2.1 Ti:Sapphire laser

For the generation of short laser pulses, a gain material with a broad gain bandwidth is required. The Ti:Sapphire crystal, which was found and demonstrated in 1986 [48], offers a gain bandwidth of about 230 nm, centered around 800 nm. Pulsed operation of such a laser oscillator requires individual modes of a cavity to be locked in phase. To achieve this in a Ti:Sapphire laser, Kerr-lens mode-locking is the technique that is most commonly used [49]. It is a passive mode-locking technique which makes use of the Kerr effect (nonlinear response of the refractive index of a medium $n(r,t) = n_0 + n_2 I(r,t)$, n_0 being the intensity independent component and n_2 being the nonlinear index of refraction) to create a fixed phase relation between adjacent longitudinal laser modes. Since the Kerr effect is a nonlinear interaction, it will be stronger for a pulse compared to a continuous wave at fixed average power. Typically the laser gain medium itself is used as Kerr-medium, where the laser pulse will create a Kerr-lens, resulting in a focusing effect in the crystal. In combination with an aperture this can lead to a higher transmission for a laser pulse than for a continuous light wave, which then can lead to a pulsed operation of the laser oscillator. This technique enabled the generation of pulses with pulse durations of less than 6 fs [50, 51]. However, pulses from an oscillator can hardly be used for high field experiment. The typical repetition rate of a Kerr-lens mode-locked Ti:Sapphire laser oscillator is in the range of 80 MHz with pulse energies in the nJ range.

These pulses, however, can be amplified in a second step. A breakthrough was the development of the chirped pulse amplification (CPA) technique [21]. This enables the amplification of nJ pulses to the mJ level at the expense of slightly longer pulse duration (around 25 fs) and reduction of the repetition rate. A pulse generated in the oscillator is sent several times through an inverted Ti:Sapphire crystal, coherently adding photons to the pulse by stimulated emission. The intensities can grow quickly and reach the damage threshold of the Ti:Sapphire crystal (approx. 10^{10} W cm^{-2}) and destroy the gain medium. If the laser pulse is stretched before amplification to some ps, while preserving its bandwidth, the intensity during amplification can be decreased dramatically. This stretching can be achieved by propagating the pulse through dispersive material or by a grating based stretcher. In both cases the different frequency components of the pulse will be delayed against each other. This preserves the full bandwidth and coherence of the pulse, while increasing the pulse duration. Typically a positive chirp is applied to the pulse (most materials have normal dispersion in this spectral range), such that the "red" part travels before the "blue" part of the pulse.

As a consequence of the spectrally dependent gain, the bandwidth of the amplified pulse will be typically reduced. This phenomenon is called gain-narrowing. An amplified pulse will have a bandwidth of roughly 30 nm which supports pulses with approximately 25 fs duration. To retrieve these short pulses from the stretched and amplified pulse, the pulses need to be recompressed. This can be done in a prism- or grating-based compressor: the frequency components are separated spatially and their geometrical path length is adjusted. Then the components are overlapped again, forming the compressed pulse. In the best situation, the compressor will compensate for the complete frequency chirp of the pulse, leading to a fully compressed pulse with a flat phase. However, this will in most cases hardly be achieved. The final pulse can be greatly improved by using an acousto-optic programmable dispersive filter, which is commercially available from Fastlite and known as "dazzler". It permits adding arbitrary phase terms to the pulse before amplification [52]. In addition it allows also for shaping the spectral amplitude. By reducing the amplitude around the gain maximum of the amplifier crystal, the obtained spectrum after amplification can be broadened. However, the pulse duration cannot be shortened significantly by this method.

To generate shorter pulses additional spectral broadening after amplification in a hollow-core fiber [53] or a filament [54] with subsequent pulse compression can be applied. This enabled the generation of pulses down to 5.1 fs with an energy in the 100 µJ regime [55].

2.2.2 The laser system used for the experiments

The laser system used for performing the experiments presented in this thesis is a Ti:Sapphire based CPA setup and is schematically shown in Fig. 2.8. The systems starts with a Kerr-lens mode-locked Ti:Sapphire oscillator from Femtolasers Produktions GmbH. It is pumped by a Verdi V-10, a diode pumped solid-state laser, delivering a continuous beam centered at 532 nm, manufactured by Coherent Inc. After compressing the output beam of the oscillator using double chirped mirrors, pulses with a duration of 10 fs at a repetition rate of 82 MHz are delivered. The beam propagates through polarizers in combination with a Pockels-Cell (PC). The PC acts as electro-optic switch and is operated at 1 kHz, synchronized to the laser oscillator. Each time it selects a single pulse to reduces the repetition rate. The beam then propagates into a commercially available all-reflective grating stretcher from Kapteyn-Murnane Laboratories, Inc. Thereby the pulses are stretched to 200 ps. Because back-reflections from the amplifier can disturb the oscillator operation, the stretched pulse is sent through a Faraday Insulator (FI). A dazzler is then used to impress additional dispersion (especially higher orders) to the pulse and to modulate the spectrum of the pulse (reduce amplitude around a wavelength of 780 nm) to achieve a short and fully compressed pulse after amplification and compression. The pulse is amplified in two amplification stages. The first stage is pumped by a pulsed actively Q-switched solid state laser (Evolution 30, intra-cavity frequency-doubled Nd:YLF laser, manufactured by Coherent Inc.). Pump pulses with a duration of 250 ns (Ti:Sapphire has a fluorescence lifetime of 3.2 μs) and a center wavelength of 527 nm with a pulse energy of 9 mJ enter the first amplification stage. The first amplifier stage is built as a 9-pass amplifier with a tri-angular geometry. Two 2" dichroic focusing mirrors (focal length 250 mm) and a large flat side mirror are used to steer the seed pulse 9 times through the amplifier crystal (Fig. 2.8). The pump beam is sent through one of the dichroic concave mirrors. The crystal is cooled down to 105 K by a closed-loop cryostat (Cryotiger, APD Cryogenics). Thereby the thermal conductivity of Ti:sapphire is increased by approximately two orders of magnitude (46 W/mK @ 293 K). This has also the advantage of decreasing thermal lensing, because the temperature coefficient of the refractive index n decreases upon decreasing the temperature. A spatial mask applied to all 9 passes reduces the amount of amplified spontaneous emission (ASE) to the µJ level. The amplified beam achieves a typical energy of 900 µJ after the first amplifier. In the second amplification stage, the beam is not focused into the laser crystal. This amplifier consists of three passes. The crystal is cryogenically cooled like the crystal of the first amplification stage. A part of the laser output of the first pump laser and an additional Evolution 30 pump laser are used to pump this crystal at a total energy of 23 mJ. The energy of the pulses is boosted to 4.6 mJ. The amplifier is well saturated, which leads to a generation of stable, high-energy pulses. A compressor based on four gratings introduces the desired negative GDD to provide 30-fs pulses with up to 3 mJ.

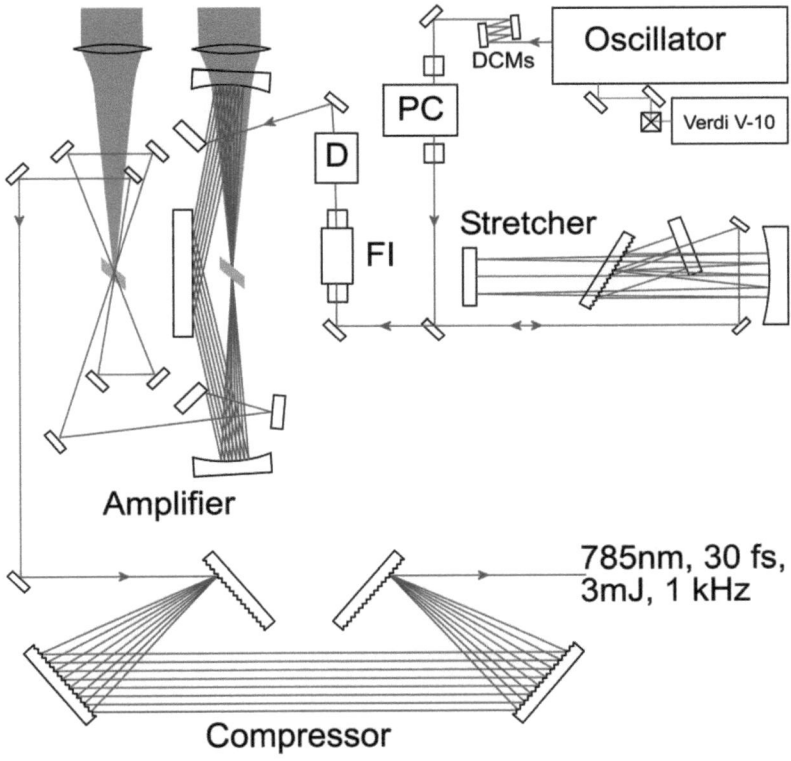

Fig. 2.8 Schematic of the laser system used in the experiments presented in this thesis. PC = Pockels cell, D = Dazzler.

Chapter 3

Quantum Path Interference

The existence of different electron quantum paths in the high-order harmonic emission is intrinsic to the high harmonic generation process. In this chapter we show that the strong field of the laser allows controlling the excursion time of these paths on an attosecond timescale. We observe a clear signature of interference between different electron quantum paths and investigate the universal properties of these quantum-path interferences (QPI) in various generation media. We discuss how ionization effects influence the observed interference structures.

3.1 Theoretical description of QPI

In the previous chapter, the underlying physics of high harmonic generation was described by the semi-classical three-step model [12, 56] in which an electron initially in the ground state is freed into the continuum by tunnel ionization, accelerated, and ultimately driven back to the core by the oscillating linearly polarized laser field where it may recombine and emit a high-energy harmonic photon (Fig. 2.3). We have seen that depending on the spectral range, several different electron trajectories can lead to recombination and contribute to the harmonic emission. Electron wave-packets following different trajectories leading to the same photon energy usually acquire a different phase (Fig. 2.6). Especially in the plateau region, the two shortest electron trajectories characterized by an excursion time shorter than one optical cycle can lead to a strong emission. The resulting phase difference between the various contributions is expected to reveal itself through interference effects. QPI is an intrinsic phenomenon of HHG and can be observed for various generation media as we will see later.

The findings based on the semi-classical model have been confirmed by a quantum-mechanical theory which has been developed within the strong field approximation (SFA) [15, 16]. Similar to the Feynman's path integral approach, the harmonic dipole moment can be written as the sum of contributions of different quantum paths. These quantum paths are a generalization of the classical electron trajectories. The phase $\phi_q^{(j)}$ associated to each quantum path (j) of harmonic order q is given by the classical action along the corresponding trajectory, and can be approximated by the product of the ponderomotive energy U_p with the electron excursion time

$$\phi_q^{(j)}(r,t) \approx -U_p \tau_q^{(j)} \approx -\alpha_q^{(j)}(I(r,t)) \cdot I(r,t).$$

In this equation, I is the intensity of the laser pulse and $\alpha_q^{(j)}$ is roughly proportional to the electron excursion time $\tau_q^{(j)}$. The longer the trajectory, the stronger the intensity dependence of the associated dipole phase. The relative phase between the different trajectories leads to interference in the total single-atom dipole moment. The dipole strength exhibits fast oscillations with increasing intensity when the harmonic is in the plateau region, where mainly two quantum paths contribute to the emission. At low intensity when the harmonic enters the cutoff, the interferences disappear because only a single quantum path contributes. The two shortest orbits, referred to as the 'short' ($j=1$) and 'long' ($j=2$) trajectories [16] have the strongest contribution to the total signal. The 'short' trajectory is characterized by a short electron excursion time $\tau_q^{(1)} < 0.65 T$, where T is the laser period. The 'long' trajectory has a longer excursion time $0.65 T < \tau_q^{(2)} < T$. This corresponds to numerical values of $\alpha_q^{(1)} \approx 1-5 \cdot 10^{-14}$ rad cm^2/W and $\alpha_q^{(2)} \approx 20-25 \cdot 10^{-14}$ rad cm^2/W [16, 37].

This allows calculating the expected modulation periodicity of the two first quantum paths (first order quantum path interferences, QPI): The intensity of harmonic order q is given by the interference of the two harmonic fields given by $E_q^{(j)}$

$$I_q = \left|E_q^{(1)} + E_q^{(2)}\right|^2 = I_q^{(1)} + I_q^{(2)} + \cos(\Delta\alpha \cdot I) \cdot \sqrt{I_q^{(1)} I_q^{(2)}}.$$

Therefore the modulation periodicity is given by

$$\Delta I = 2\pi / \Delta\alpha \approx 0.3 \cdot 10^{14} \text{ W/cm}^2.$$

Here $I_q^{(j)}$ denotes the intensity of the harmonics and I is the intensity of the laser pulse. This also means that these oscillations are smeared out if integrated over laser intensities varying by a small amount in the order of $2\pi / \Delta\alpha$.

3.2 Experimental conditions for the observation of QPI

In the experimental condition where the gas jet is positioned in the focus of an ultrashort laser beam, the spatial and temporal averaging due to the spatial and temporal beam profile will smear out the QPI oscillations in the macroscopic response. Indeed only monotonic, unmodulated distributions have been measured previously [57, 58]. While the contributions from the two main quantum paths have been independently characterized experimentally [17, 18, 59-61], their relative phase has not been characterized before. In this thesis experimental conditions have been found that permit for a direct observation of QPI.

In order to avoid temporal averaging, we perform spectral filtering, while spatial averaging is partly avoided through far-field spatial filtering. The latter also allows one to balance the weight of the contributions of the two quantum paths in the detection to increase the interference contrast. The measured contrast is high enough to resolve the QPI over all generated plateau harmonics. This gives access to the relative phase of the first two quantum paths. This is a first step towards a reconstruction of the single-atom dipole moment. It gives insight into the ultrafast electron dynamics in the emission process. Moreover, by slightly changing the laser intensity, we demonstrate a control of these paths on an attosecond timescale. Indeed, only tiny changes (of a few 10 as) of the relative excursion times of the electrons are enough to shift their relative phase by π.

3.2.1 Temporal averaging

Let us first consider temporal averaging effects due to the envelope of the laser pulse. The harmonic chirp is given by

$$\Delta\omega^{(j)}(t) = -\frac{\partial \phi_q^{(j)}(t)}{\partial t} = \frac{\partial \alpha_q^{(j)}(I(t))}{\partial I} \frac{\partial I(t)}{\partial t} \cdot I(t) + \alpha_q^{(j)}(I(t)) \cdot \frac{\partial I(t)}{\partial t} \approx \alpha_q^{(j)} \cdot \frac{\partial I(t)}{\partial t},$$

where $\Delta\omega = \omega - \omega_q$ [8, 46, 62, 63]. The harmonic chirp spectrally shifts the harmonic emission generated on the leading edge to higher frequencies and the emission of the trailing edge to lower frequencies (Fig. 3.1). The last conversion assumes a weak intensity dependence of α, which is true for harmonics in the plateau region (see Fig. 3 in [37]). By limiting the observation of the interferences to frequencies close to the exact harmonic frequency we avoid temporal averaging. The harmonic chirp ensures that the generation of this center frequency is always confined to the instant of maximum laser peak intensity in a Gaussian or similarly bell-shaped pulse (indicated by the dashed black lines in Fig. 3.1).

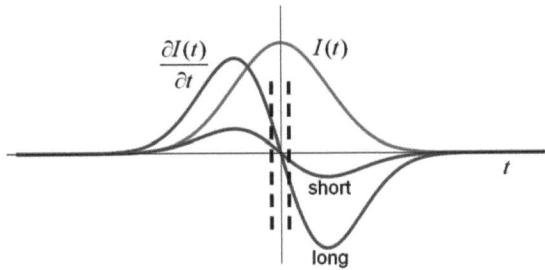

Fig. 3.1 The intrinsic harmonic chirp (blue, arb. u.) spectrally shifts the harmonic emission generated at different instants in the laser pulse (red, arb. u.). The chirp is different for the long and the short trajectory (see also [63]). The central frequency of the harmonic emission is for both trajectories generated at the peak intensity of the laser pulse.

The chirp, and thus the spectral bandwidth of the harmonics, increases with laser intensity. The emission generated on the edges of the envelope of the driving laser pulse is shifted to different side frequencies (i.e., the harmonic chirp effectively maps time to frequency $\Delta\omega = \omega - \omega_q$). By spectrally resolving the harmonics, we thus get rid of the temporal averaging. Even if the chirp for the contribution of the long trajectory is larger than of the short one, the spectral bandwidth where the two contributions overlap is generally large enough to allow for the detection of modulations of the harmonic yield. Detailed numerical calculations are presented in [64]. The results confirm our observations and highlight the robustness of the interference effects.

3.2.2 Spatial averaging

After discussing temporal averaging effects we can turn to the spatial domain. Averaging along the propagation direction can be reduced by using a generating medium shorter than the confocal parameter of the laser. As for the transverse averaging, we may use the spatial analog of the temporal or spectral filtering. The intensity dependence of the dipole phase leads to a curvature of the harmonic phase front (equivalent to a spatial chirp) [24, 46, 63]: the (near-field) distribution at the exit of the medium is "projected" into the far field so that a far-field spatial selection can prevent transverse spatial averaging. However, an additional difficulty arises here due to the different phase matching of the two quantum paths (see Sec. 2.1.4). We obviously need generating conditions where both paths give significant contributions to the macroscopic response. This implies focusing the laser after the generating medium. In this situation, the short path is phase matched on axis, and the long path off axis. This results in a concentration of the contributions from the short path at the center, and from the long path at the outskirts of the far-field profile. The relative contribution of the two trajectories thus varies radially, and there is an optimum position (off axis) where the fringe contrast of QPI will be maximized. A far-field spatial filter positioned at this place will in addition limit the spatial averaging. The importance of this spatial selection for the observation of QPI is demonstrated experimentally in Sec. 3.4. It is confirmed by simulations of the full macroscopic response calculating the single atom response using the dipole of SFA and a three-dimensional propagation code that solves the propagation equations for the laser and harmonic fields in the paraxial and adiabatic approximation. All simulations shown in this chapter have been performed by T. Auguste, J. P. Caumes and P. Salières. Fig. 3.2 shows the results of the simulation for harmonic 15 as a function of laser peak intensity with and without the spatial selection. In the latter case (left image), the integration is performed over the total spatial profile, but the main contribution comes from the 6-mrad central part (the signal is strongest on-axis). The spatial averaging results in blurring the QPI and therefore the harmonic amplitude is not modulated, but the spectral width still increases with the peak laser intensity as expected. In contrast, when the far-field selection is performed (right image), the harmonic spectrum reveals a clear parabolic shape.

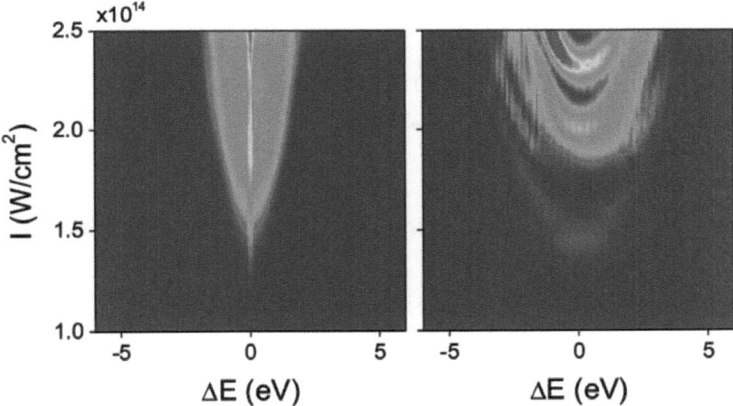

Fig. 3.2 Calculated macroscopic spectra of the 15th harmonic generated in argon as a function of laser peak intensity without (left) and with (right) a far-field 6-mrad off-axis window. The harmonic signal is plotted with a linear color scale defined as blue at zero strength to red at maximum strength (10^{10} without window and $4.5 \cdot 10^7$ with window). By courtesy of T. Auguste.

The fringe contrast is maximum at the central frequency of the harmonic ($\Delta E = \hbar \Delta \omega = 0$), as expected for first-order QPI. The fringes are shifted to higher laser peak intensity when $|\Delta E|$ increases because these frequencies are generated on the temporal edges of the laser envelope and thus correspond to a smaller effective intensity. This explains the parabolic shape of the QPI. The harmonic signal is reduced by 1 to 2 orders of magnitude when the spatial selection is performed, which makes the experimental observation of QPI challenging.

3.3 Experimental setup

In the experiments, we used neon, argon and xenon as generation media. In all gases we were able to observe QPI. Laser pulses with a center wavelength of 800 nm and a duration of 30 fs are sent through an iris aperture and are focused by a spherical mirror into a pulsed gas jet. A schematic of the experimental setup is shown in Fig. 3.3.

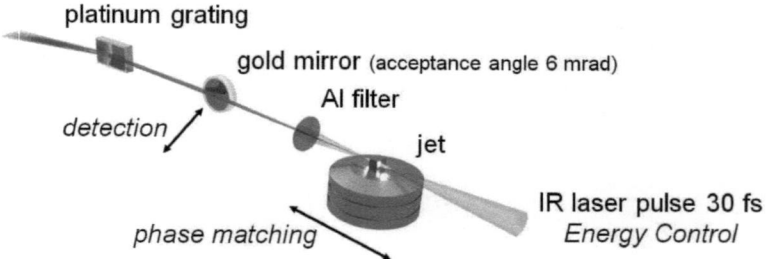

Fig. 3.3 Schematic experimental setup. A detailed description is given in the text.

The ROC of the focusing mirror was 500 mm for the case of argon and xenon, and 250 mm for the case of neon. The gas jet has a small hole, where the gas expands freely into the generation area. The diameter of the hole is 500 µm only, which creates the shortest interaction length possible in our setup. This will reduce longitudinal averaging as discussed in section 3.2.2. A motorized half-wave plate and a polarizer are used to control the energy of the laser pulses. High energy stability of the laser is required to ensure that QPI is not smeared out due to intensity fluctuations. This is achieved by operating the laser system in a well saturated regime (see Sec. 2.2.2). The jet is movable along the propagation direction to control phase matching conditions. The emitted harmonic radiation is refocused by a spherical gold mirror (ROC=30 m, size 2") in grazing incidence (approx. 87 degrees). This results in an excellent reflectivity of more than 84% in the spectral range of interest. This mirror is movable perpendicular to the propagation direction for lateral windowing of the beam. The refocused harmonics pass an aluminum filter (150 nm thickness) to remove low order harmonics from the beam, including the fundamental frequency. This is necessary because the detector used is sensitive in the IR spectral range. The harmonics pass the entrance slit of an XUV spectrometer consisting of a reflective grating and a XUV CCD camera.

Because of a strong absorption of high harmonics in air, experiments using high harmonics have to be performed in vacuum (for more details see Sec. 4.2). The experimental scheme depicted in Fig. 3.3 was implemented in the "first generation" high harmonic beamline that was built at ETH Zurich in 2001. It is shown in Fig. 3.4.

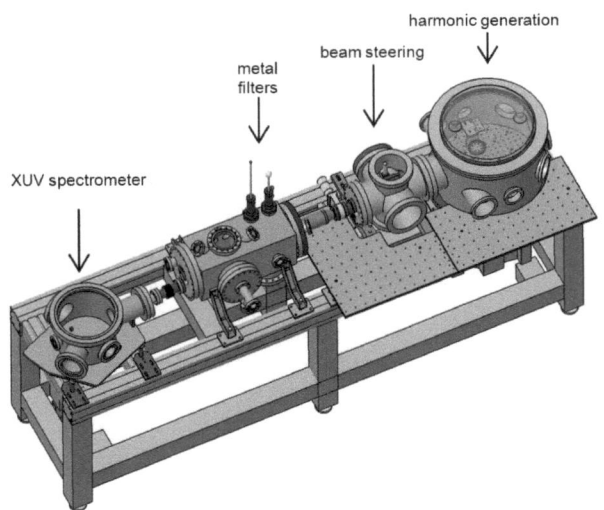

Fig. 3.4 High harmonic beamline used to perform the experiments on QPI.

3.4 Experimental results

A generic behavior with respect to the phase matching conditions, independent of the ionization potential of the generating medium, is observed. When selecting the short trajectory (jet after the laser focus), we observe spectrally narrow harmonics as a result of the small frequency chirp of the short trajectory. Their amplitude increases monotonically with the laser intensity since in this case only one trajectory is detected. With the experimental setup described above, we can filter out transverse sections of the emitted beam in its far field distribution. Changing the far field spatial selection from on-axis to off-axis with phase matching optimized for the short trajectory only affects the overall detected harmonic yield, reducing it by roughly one order of magnitude. This behavior is experimentally shown for the case of argon in Fig. 3.5.

In contrast to this situation, both trajectories are phase-matched for a jet placed before the laser focus. For an on-axis spatial selection in the far field, the harmonic signal is still spectrally narrow because the strong emission from the short trajectory with its lower divergence is detected mainly. When selecting an off-axis portion of the beam, a clear broadening of the spectral width of the plateau harmonics with increasing laser intensity is observed (see Fig. 3.6). This is the typical signature of the long trajectory resulting from its larger intrinsic chirp. When properly balancing the relative contributions from the short and the long trajectories by adjusting the position of the far-field spatial filter the contrast of the interference is maximized and a clear modulation of the harmonic yield with laser intensity occurs. This interpretation is further supported by the fact that in the cutoff region we neither observe significant broadening nor any modulations since only one quantum path exists.

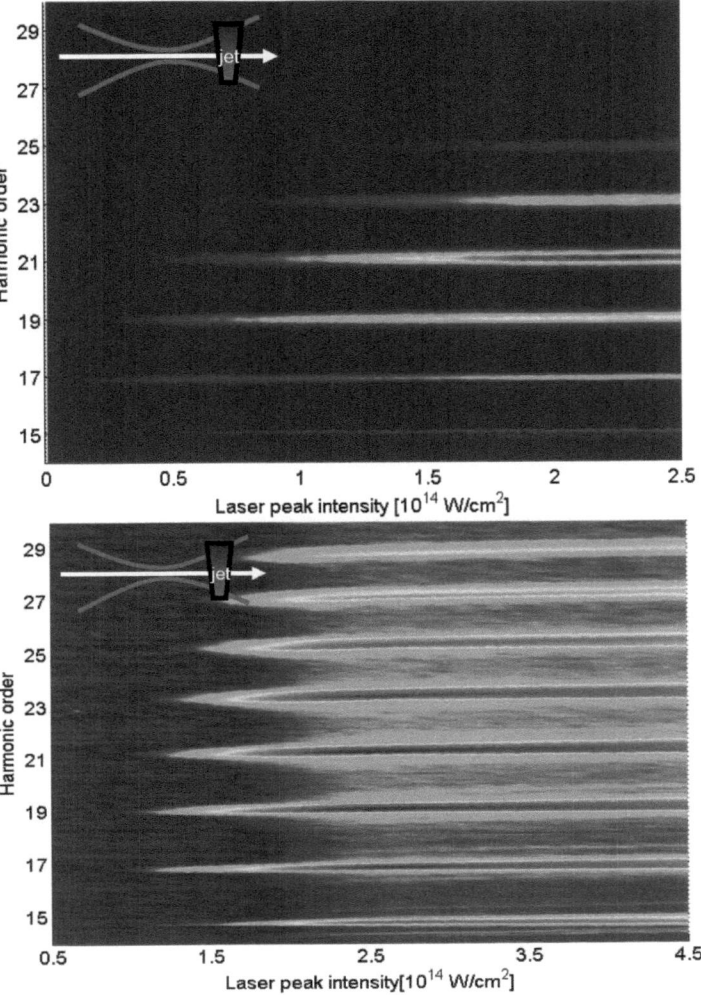

Fig. 3.5 Spectral intensity of high harmonics versus laser peak intensity generated in argon. The jet has been placed after the laser focus, preferably selecting the short trajectory in the phase matching process. In the upper image, the spatial selection was positioned on-axis, on the lower image it was off-axis.

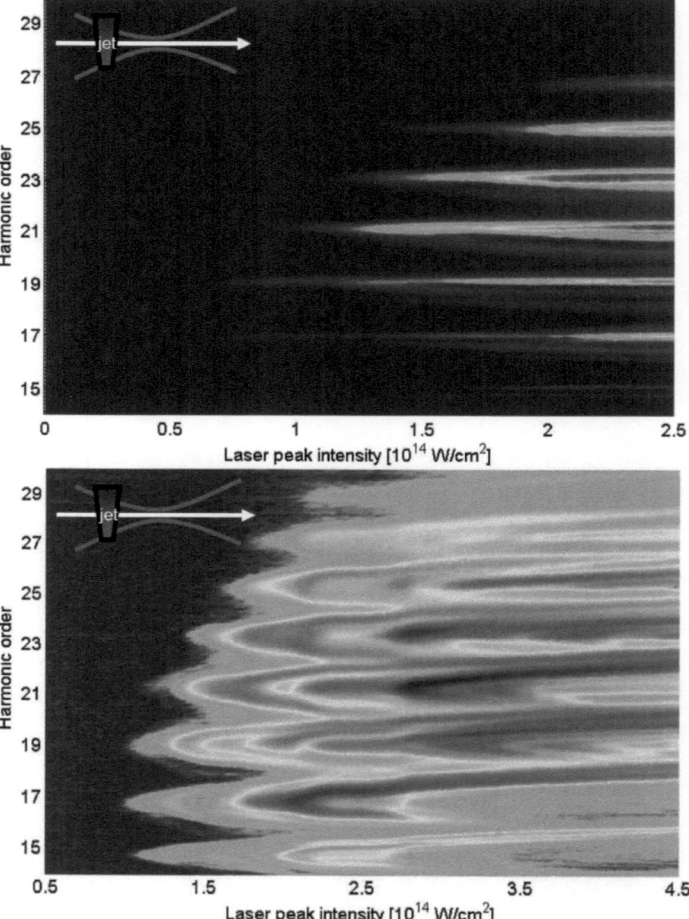

Fig. 3.6 Spectral intensity of high harmonics versus laser peak intensity generated in argon. The jet has been placed before the laser focus, the short and the long trajectory are both phase matched. In the upper image, the spatial selection was positioned on-axis, on the lower image it was off-axis. Compared to the on-axis selection, the long trajectory with its larger divergence is detected in the case of an off-axis selection which leads to a spectral broadening of the individual harmonics.

3.4.1 Argon

The harmonic spectra measured in argon in the optimized conditions for QPI observation are shown in Fig. 3.7 (a). The intensity for barrier suppression in argon (ionization potential 15.8 eV) is $I_{BS} = 2.5 \cdot 10^{14}$ W/cm^2 (see Sec. 2.1.1). Below this intensity, we observe several modulation periods corresponding to QPI as shown in Fig. 3.8. For higher intensities, the onset of ionization effects results in both an asymmetric shape of the different harmonic orders and a much weaker modulation of the harmonic signal. These trends are reproduced in our simulations, as for harmonic 21 shown in Fig. 3.7 (b) (6 mrad wide far-field integration window centered 17 mrad off axis). The experimentally measured average modulation periodicity $\approx 3 \cdot 10^{13}$ W/cm^2 is consistent with the expected value $2\pi / \Delta\alpha$.

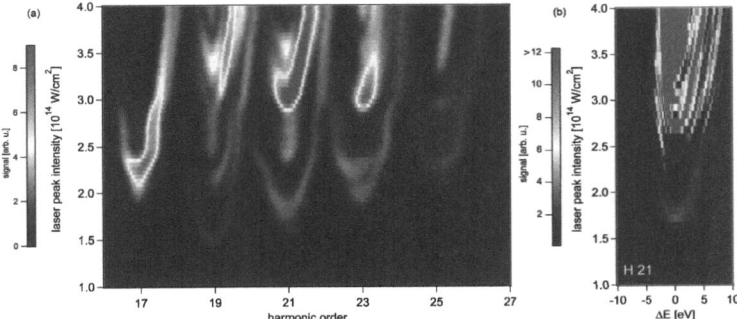

Fig. 3.7 Harmonic spectra generated in argon versus laser peak intensity with short and long trajectory phase-matched and off-axis spatial selection: (a) measured spectra, (b) simulated spectra around harmonic 21, courtesy of T. Auguste.

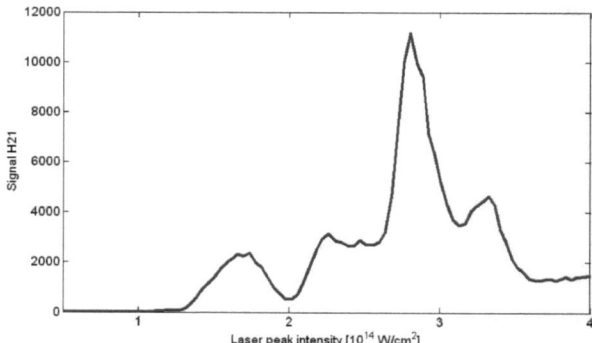

Fig. 3.8 Spectrally integrated signal in a narrow window at harmonic order 21 from Fig. 3.7 (a) showing several modulations of the harmonic yield due to QPI.

We further analyzed the origin of the different fringe structures with detailed simulations. Fig. 3.9 shows the result of the full model compared to simulations with either all ionization effects switched off or only the single-atom level ionization effects included. The latter case thus includes the depletion of the medium but does not take into account macroscopic effects of free-electron dispersion (i.e. change of the refractive index of the medium). The simulations show that the asymmetry of the harmonic spectrum at high intensity results from ground state depletion, which is a single atom effect. The harmonic signal on the red side of the spectrum is reduced because on the trailing edge of the generating laser pulse the medium has already been depleted. The asymmetry of the harmonic spectrum is thus a direct result of the time-to-frequency mapping caused by the harmonic chirp. The simulated data in Fig. 3.9 also reveal that the apparent fringe splitting on the blue side of the exact harmonic frequency and other distortions of the fringe structure are a direct consequence of free-electron dispersion. These free electrons are created on the leading edge of the pulse for intensities beyond barrier suppression. Since the free electrons are concentrated on the beam axis, it is expected that they affect the emission from the short trajectory more strongly than from the long trajectory because of their different phase-matching properties [15, 63]. Interference in the far field between contributions from the short and long trajectory originating from different transverse locations in the medium explains the observed distortions in their relative phase.

Different particle densities in the interaction region do not affect the described spectral behavior significantly in our measurements as well as in our simulations. In particular the modulation periodicity remains unchanged, indicating that dispersion does not play a dominant role for the modulations of the harmonic yield. However, closer inspection reveals subtle changes in the fine structure of the fringes when the pressure is increased. They appear only for the high-pressure scenario and above the barrier suppression intensity. Therefore we attribute these changes to macroscopic nonlinear optical effects.

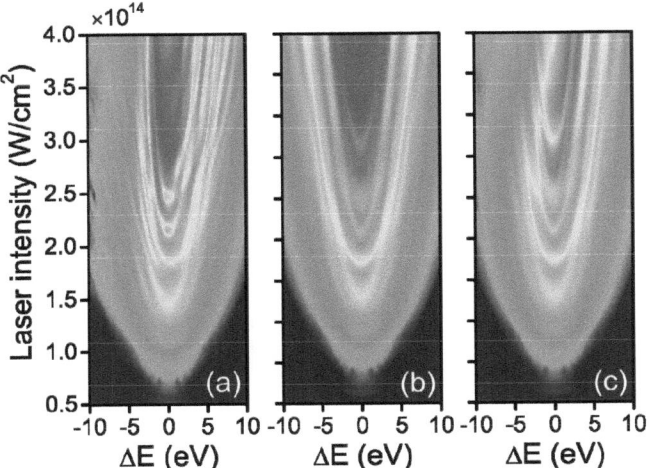

Fig. 3.9 Simulated QPI for H15 in argon at a pressure of 10 Torr. (a) full simulation, (b) ionization switched off, (c) with depletion, but free electron dispersion switched off. By courtesy of T. Auguste.

3.4.2 Xenon

Xenon with its lower ionization potential of 12.1 eV shows saturation effects at lower intensities than argon. The intensity for barrier suppression is $I_{BS} = 8.7 \cdot 10^{13}$ W/cm^2. Above this intensity, the harmonic yield saturates and the effective pulse duration of the laser pulse contributing to the harmonic emission gets shorter with increasing laser intensity due to the fast ionization of the medium. Only the leading part of the pulse generates harmonics, resulting in a strong asymmetry of the harmonic spectra. Meanwhile, the harmonic chirp increases further, which leads to asymmetric broadening on the blue side only (Fig. 3.10).

Still, one modulation of the harmonic yield at the exact harmonic frequency can be measured under these conditions as shown in Fig. 3.11 for harmonic order 17.

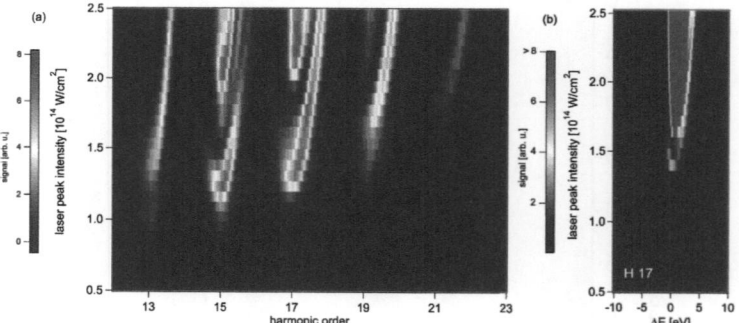

Fig. 3.10 (a) Measured harmonic spectra generated in xenon versus laser peak intensity, (b) simulated spectrum for harmonic order 17, courtesy of T. Auguste.

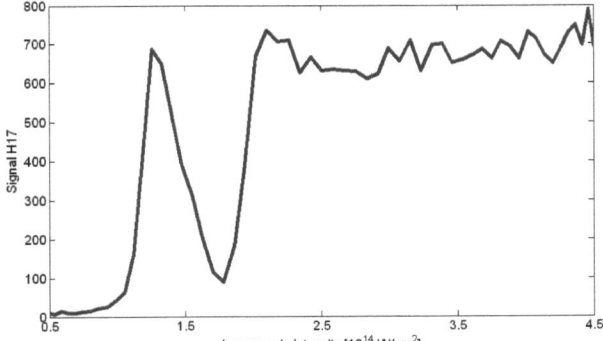

Fig. 3.11 Spectrally integrated signal in a narrow window at harmonic order 17 from Fig. 3.10 (a) showing the modulation due to QPI. Only one period can be observed due to fast depletion of the medium.

3.4.3 Neon

In the case of neon, the high ionization potential of 21.6 eV leads to a BSI of $I_{BS} = 8.3 \cdot 10^{14}$ W/cm^2, which is approximately one order of magnitude higher than in the case of xenon. Therefore it is possible to generate high harmonics efficiently well below BSI, opening the way to study the QPI over a wide range of intensities without being limited by ionization. As expected, the data displayed in Fig. 3.12 shows no sign of blue shift.

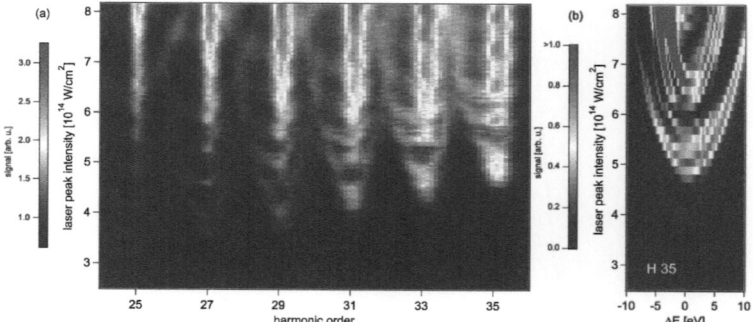

Fig. 3.12 (a) Experimental harmonic spectra generated in neon with respect to the laser peak intensity: over the full range of accessible intensities located below ionization threshold neither a blue shift nor a saturation of the harmonic signal at the harmonic central frequency is observed, (b) simulated spectrum for harmonic order 35, courtesy of T. Auguste.

Within the range of covered intensities, we are able to clearly identify about five modulation periods of the harmonic yield of harmonic order 35 with an average periodicity of $\sim 8 \cdot 10^{13}$ W/cm^2 which is shown in Fig. 3.13.

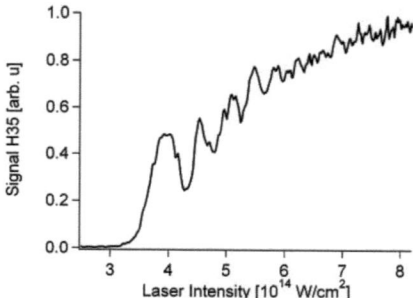

Fig. 3.13 Spectrally integrated signal in a narrow window at harmonic order 35 from Fig. 3.12 (a) showing the modulation due to QPI with a modulation periodicity of $\sim 8 \cdot 10^{13}$ W/cm^2.

3.5 Influence of CEP, phase-matching and phase modulation on QPI

At a sufficiently high intensity, neighboring harmonics start to overlap, which is clearly visible in the case of argon (Fig. 3.7). If the laser carrier envelope offset phase (CEP), i.e. the position of the electric field under the laser pulse envelope, was stabilized [65-67], one would observe interferences between long path contributions from neighboring orders generated at different times of the laser envelope [18]. In our experiments, these interferences were blurred out and did not perturb the observation of QPI because the CEP was random and multicycle laser pulses have been used. QPI does not depend on the laser CEP since it occurs on the subcycle attosecond timescale (within a single laser cycle). The amplitudes of the plateau harmonics are clearly modulated as a function of laser peak intensity.

Changes in phase matching may also occur with increased intensity due to an increased density of free electrons, but they are not expected at lower intensities where the structures of QPI first appear.

Finally, the strong phase modulation of the long trajectory alone may produce spectral interferences due to the emission of the same frequency at two different instants of the temporal envelope of the laser (see also Fig. 3.1). This corresponds to a spectral redistribution of energy. Therefore, the spectrally integrated harmonic signal (integrated over a complete single harmonic) would not show any modulation upon varying the laser peak intensity. This is in contrast to our measured and simulated spectra which still exhibit modulations in this case. Furthermore, at the exact harmonic frequency, the above-mentioned spectral interferences do not occur, since emission of this frequency happens at a single instant in time only (see also Fig. 3.1).

3.6 Spatially resolved QPI

In a follow-up experiment we improved the spatial filter, which is required to observe QPI. Instead of making a spatial selection using a movable mirror, we directly use the entrance slit of a XUV spectrometer as spatial filter. The harmonic beam is misaligned in the direction of the slit, which enables the suppression of the strong emission of the short trajectory. Thereby one can balance the ratio of the emission from the short and the long trajectory. The spectrometer used in this case was not based on a XUV CCD camera, but had a MCP/phosphor screen combination. The images were then captured by a CCD. The contrast of this detection scheme is worse compared to a XUV CCD and limited our measurements to argon gas. In Fig. 3.14 a schematic of the experimental setup is shown. In the inset the misaligned beam is indicated.

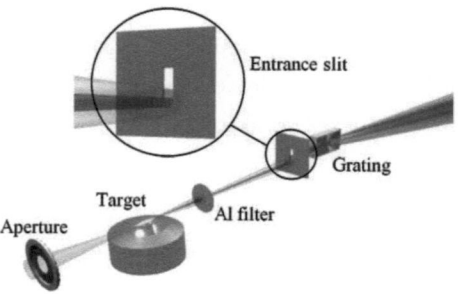

Fig. 3.14 Scheme of the experimental setup for the observation of spatially resolved QPI. The strong emission of the short trajectory is indicated in dark blue. It is surrounded by its weaker part plus the emission from the long trajectory. The beam was misaligned on the entrance slit to balance the contributions from the two trajectories. This enabled the observation of spatially resolved QPI.

The measurement of spatially resolved QPI enabled us to fully reproduce the results obtained by the spatially integrated measurements that have been presented in the previous sections. This has been achieved by performing the spatial integration numerically. In Fig. 3.15, a spatially resolved interference structure is shown for harmonic order 19 and 21. A clear parabolic structure is visible. The spatial interference behavior can be explained by the different curvature of the harmonic wavefronts generated by the short and the long trajectory. The spectral structure can (as before) be explained by the time-to-frequency mapping of the harmonic chirp and the temporal structure of the generating pulse.

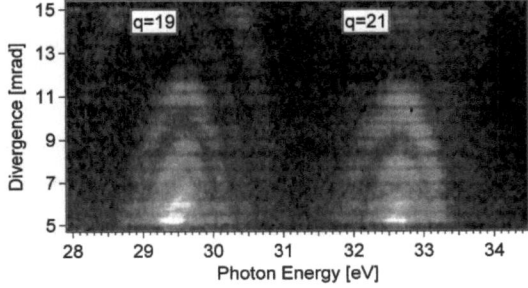

Fig. 3.15 Spatially resolved interference structure measured at a fixed peak intensity of $3.4 \cdot 10^{14}$ W cm^{-2} for harmonic order 19 and 21 generated in argon.

3.7 Conclusion

The presented data show that experimental and theoretical conditions can be found enabling the observation of quantum path interferences in high-harmonic generation independent of the harmonic generating medium. This is made possible by means of proper spectral and spatial filtering. The filtering prevents spatial and temporal variations of the interference conditions from smearing out the fringes. The influence of the generating medium, specifically ionization effects, on the observed interference structures were discussed. The comparison of measurements taken in different gases allowed us to study the behavior of QPI with laser intensities above, around and below BSI. Depending on the regime, the plateau harmonics exhibit intensity-dependent features such as saturation of the harmonic yield, spectral blue shift, and trajectory dependent spectral broadening. The measurements are well supported through simulations performed by T. Auguste, J. P. Caumes and P. Salières. Our method provides the possibility of investigating this behavior in more complex systems such as molecules or clusters. Recent experiments have shown that in small molecules, such as CO_2, N_2, H_2, D_2, the continuum dynamics of the electron wavepacket (which is at the origin of QPI) is similar to that of atoms, at least for the short trajectories [68, 69]. One can thus expect that QPI will be observable in these systems and that its interferometric sensitivity will reveal differences between the molecular and the atomic continua. Our method is a general, robust tool to collect information directly in the spectral domain of the harmonic generation process including its temporal dynamics under varying experimental conditions. This interference is not CEP sensitive since it occurs on the subcycle timescale of the quantum paths. By varying the laser intensity, we change their relative phase and thus demonstrate a path control on an attosecond timescale.

Chapter 4

A New Beamline For Attoscience

The experiments described in the previous chapter on QPI were performed in the first generation high harmonic setup which was built at ETH in 2001 (see Fig. 3.4). This setup enabled measurements in the spectral domain. As described in section 2.1, high harmonic generation is a coherent up-conversion process. There is a fixed phase between the harmonics. During each half-cycle of the driving field an XUV pulse with a duration in the attosecond regime can be emitted. In case of a multicycle driving field the continuous harmonic spectrum will become discrete (odd harmonics) and the emission can form an attosecond pulse train in time domain (see Sec. 2.1.5). The temporal and spatial coherence [24] make high harmonics an interesting light source in the XUV spectral domain. The pulses allow for pump-probe measurements with high temporal resolution.

Due to the low generation efficiency of the high harmonics (typically in the range of 10^{-5} to 10^{-8}), the reachable intensities in the XUV will typically not induce nonlinear processes in a target medium. It has been shown that it is possible to measure a second-order autocorrelation of high harmonics with attosecond resolution in helium [22] where the harmonics had photon energies below the ionization potential of helium. For harmonics with higher energy, the medium is ionized by a single photon process which makes it difficult to extend this autocorrelation technique further to even shorter wavelengths: the relevant two-photon cross-section becomes too small [23]. This makes XUV pump - XUV probe measurements challenging.

It is possible to use a replica of the driving electric field as probe beam, where in contrast to the harmonics, high pulse energies are available. The attosecond pulses in a pulse train are synchronized to this field in time due to the generation process (see Fig. 4.1).

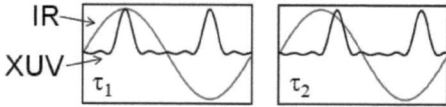

Fig. 4.1 Typical XUV/IR pump-probe measurement shown for two different time delays τ_1 and τ_2. The attosecond pulses (labeled with XUV) are synchronized to the driving IR field (IR).

Experiments of this type are only possible if the experimental setup offers interferometric stability. In addition, an interaction target for the attosecond pump-probe measurements needs to be available. This target medium will typically be ionized due to the high photon energy in the XUV spectral region. The measurement of a photo-electron or –ion spectrum can therefore be a suitable detection scheme in an attosecond experiment.

The first generation high harmonic setup at ETH Zurich didn't fulfill the requirements for attosecond time-resolved measurements which lead to the decision of designing and building a new high harmonic beamline. In this chapter a detailed technical description of the construction will be given. The new setup was applied in various experiments with attosecond time resolution. They are described in chapter 5 and 6.

4.1 Basic requirements and optical layout

The most important requirements for the new beamline are

- interferometric stability in optical setups (better than $(\lambda/2)/10 = 40$ nm rms jitter in delay lines, corresponding to 133 as, λ being the wavelength of the driving field)

- high flexibility: the beamline should be very versatile such that it can be used for different future experiments

- the existence of an interaction region (with a focal spot), where the attosecond experiment takes place, equipped with a suitable photoion or photoelectron detection.

The new beamline is installed completely on an optical table, which itself offers interferometric stability. The basic idea is to transfer this stability into the vacuum chambers of the beamline. The simplest optical layout requires one focal spot for the generation of the attosecond pulses. This spot then should be imaged to a second focus, where the attosecond experiments take place. We used atomic gases as interaction medium for these experiments. Most of the photons are transmitted through this gas target. Therefore it is possible to install an XUV spectrometer after the interaction region. This allows a simultaneous detection of photoelectrons/ions and photons, which is useful for optimizing and monitoring the harmonic emission. In our case, the acquisition of a complete XUV photon spectrum takes some hundred milliseconds only, whereas measuring a photoelectron spectrum can take several minutes. More details on this will be given at a later point.

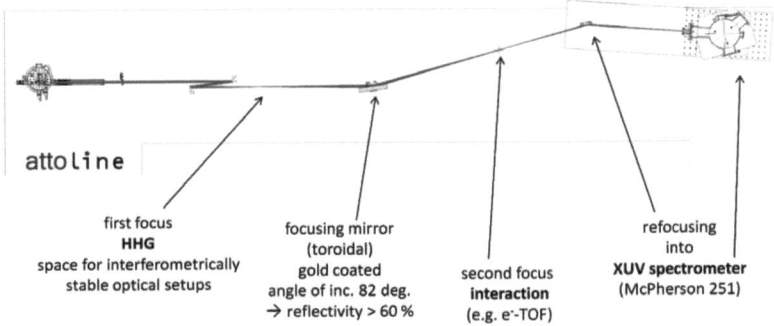

Fig. 4.2 Basic optical layout of the new high harmonic beamline.

The basic optical layout is shown in Fig. 4.2. The laser beam enters from left and is focused by a spherical mirror into the generation target. After some propagation, the beam is refocused by a gold coated toroidal mirror into the interaction region. This type of mirror is chosen because of the following reasons:

A high and broadband reflectivity of the optics is required for highest flexibility concerning possible future experiments. A metal surface at normal incidence has a reflectivity of approximately 5% in the harmonic wavelength regime. Increasing the angle of incidence increases the reflectivity according to the formulas of Fresnel [70]. (Note that the refractive index of most materials is smaller than one in the XUV. In principle this allows for total internal reflection for very large angles). The required size of the mirror thereby increases, which sets an upper limit to the angle of incidence. We have chosen an incidence angle of 82 degrees and a mirror size of 255x60 mm². Gold was selected as mirror coating (thickness 40 nm), resulting in a reflectivity > 60% in the wavelength range of the high harmonics. In Fig. 4.3 the reflectivity of our mirror versus photon energy in the typical wavelength range of the harmonics is shown for p-polarized light [71].

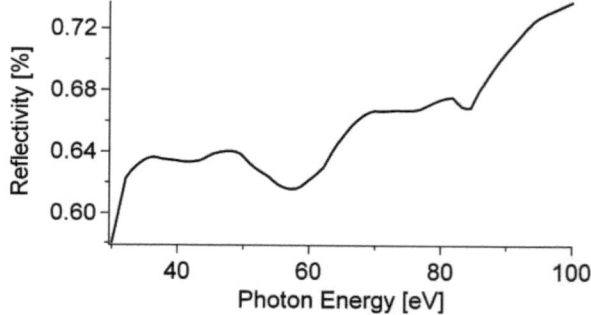

Fig. 4.3 Reflectivity of a gold surface at an angle of incidence of 82 degrees for p-polarized light in the typical wavelength range of the high harmonics [71].

Now that the material properties of the mirror are selected, the mirror geometry has to be chosen. The best point-to-point imaging is achieved by an ellipsoidal mirror geometry. However, raytracing showed that this geometry is very sensitive to potential misalignment and might be very challenging from an experimental point of view. The situation can be relaxed by using a toroidal mirror geometry. Thereby the section of the ellipsoid that forms the mirror is approximated by constant radii in meridial and sagital directions. The disadvantage of a toroidal mirror is that it has intrinsic aberrations. Time smearing of the reflected radiation can occur and has to be investigated carefully. The complete ray tracing of the setup has been done using ShadowVUI, which is a part of the X.O.P. (X-ray Oriented Programs) package [72]. The source-to-mirror distance and mirror to target distance is chosen to be 118.6 cm. For this case the required meridial radius is calculated to be 8529.6 mm, the sagital radius is 165.2 mm. Since the wavelength of the harmonics is only a fraction of the driving laser wavelength, the surface quality of this mirror needs to be very good. For harmonic order 51 the wavelength is approximately 15 nm. Therefore, the surface roughness is specified to be better than 0.5 nm rms. The superpolished mirror was manufactured and coated by Carl Zeiss Laseroptik. They finally characterized the surface roughness by an interferometric measurement to be better than 0.32 nm rms.

As already mentioned, the radiation that is not absorbed in the interaction region is measured in the XUV spectrometer. To increase the flux in the spectrometer, the beam is refocused after the interaction focal spot onto the entrance slit of the XUV spectrometer by a spherical mirror under grazing incidence. A standard 2 inch substrate (surface roughness ~ 65 nm) was used in this case, because aberrations in the wave-front of the harmonics are not important for the spectral detection. The coating is chosen to be gold as for the toroidal mirror. The distance from the interaction region to the mirror is 800 mm and the distance to the spectrometer is 940 mm. The angle of incidence is 80 degrees. The radius of curvature of this mirror is chosen to be the meridial radius of a toroidal configuration at this position. It is calculated to be 500 mm. The spherical geometry creates a focus of the beam in one direction at the desired distance. An astigmatic line-shaped focus on the entrance slit of the spectrometer is formed. The entrance slit has a width of 100 μm. In Fig. 4.4 the simulated beam profiles for the generation focal spot, the interaction focal spot as well as the astigmatic focus for the spectrometer are shown. When starting with a focus with a diameter of 60 μm in the generation, we will create a focal spot of 60.6 μm in the interaction region. The line width at the entrance slit of the spectrometer will be 77 μm.

Fig. 4.4 Raytracing performed using ShadowVUI [72]. Generation focal spot (beam diameter 60 μm, left); interaction focal spot (beam diameter 60.6 μm, center); beam profile on the entrance slit of the spectrometer (beam height 1.94 mm, beam width 77 μm, right).

4.2 Overview of the vacuum system

Before discussing the individual sections of the setup in detail, some general remarks on the vacuum properties and the generation of the vacuum conditions will be given. Good vacuum conditions are required for experiments in the XUV spectral region. At atmospheric pressure and for a propagation distance of 30 μm half of the harmonic photons (with an energy of 50 eV) are already absorbed in air. In contrast, this distance is increased to 30 m at a pressure of 10^{-3} mbar [71].

When operating the gas targets, the pressures in the setup range from 10^{-3} mbar (generation part) down to 10^{-6} mbar (interaction part) and are achieved using turbomolecular pumps and differential pumping stages between individual vacuum chambers. A high pumping speed (in our case more than 3000 l/s) is necessary due to the high leak rates of generation and interaction gas targets. To minimize transfer of vibrations generated by the turbo pumps to the vacuum chambers, dampers have been developed. Vibration measurements using an accelerometer (ADXL330, Analog Devices) show a reduction of the vibration at the frequency of the turbo pump of about 25 dB. The dampers reduce the pump capacity of the pump by 8% typically. They are made up by flexible bellows with an inner diameter equal to the diameter of the turbo pump. The flexible bellow is fixed by four holders which are welded to each end. The two ends are then connected by two rubber dampers (Rundpuffer Typ C, NR43 Shore A, Angst+Pfister AG). An image of a complete damper is shown in Fig. 4.5. The dampers offer enough stability and enable mounting of most pumps with any orientation without additional support of the pump. They are used for all turbo pumps of the setup.

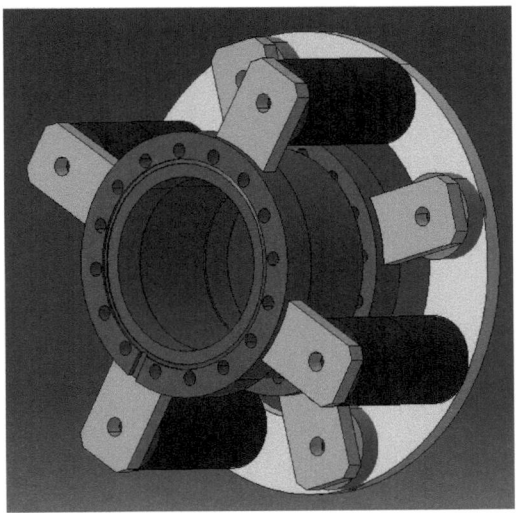

Fig. 4.5 Damper for turbomolecular pumps, reducing the vibrations transferred to the setup by 25 dB (at the rotation frequency of the turbopump). The two flanges are only connected via a flexible bellow which is stabilized by rubber dampers.

The vacuum chambers were designed around the optical layout shown in the previous section (Fig. 4.2). Enough space for optical setups surrounding the generation focal spot should be available. The beamline can be divided into four parts: 1) generation chambers, 2) diagnostics and refocusing chamber, 3) interaction chamber, 4) refocusing and XUV spectrometer. Each part will be discussed separately in the following sections. An overview of the setup is shown in Fig. 4.6.

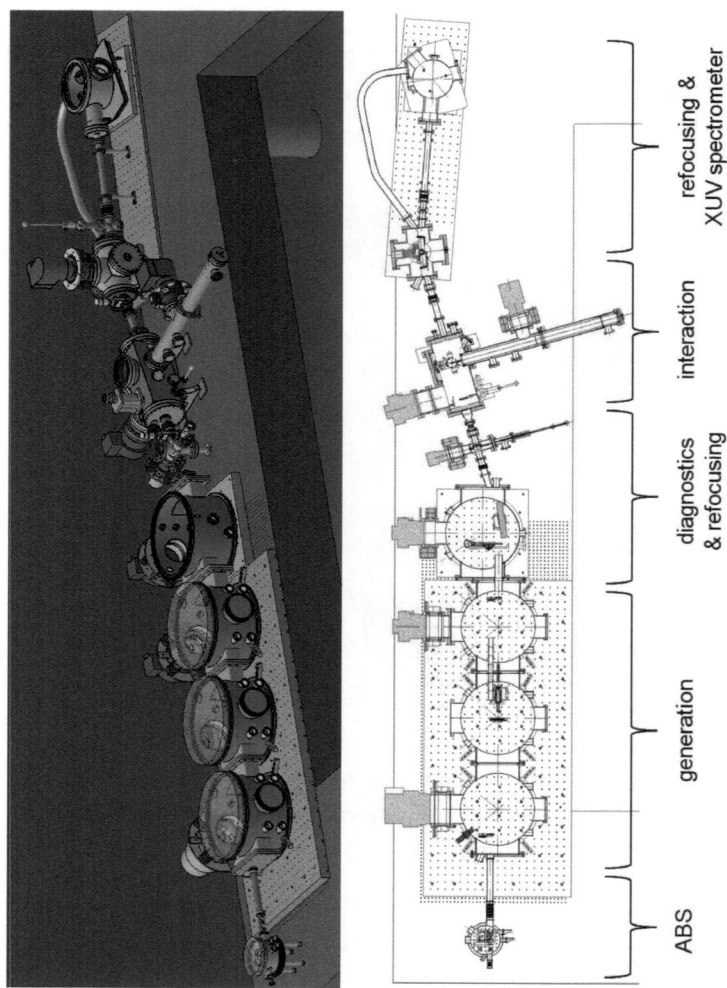

Fig. 4.6 Overview of the harmonic beamline and its vacuum chambers. (ABS = active beam stabilization)

4.3 Generation chambers

Interferometric stability is the most important requirement in this part of the setup. To transfer the stability of the optical table to vacuum, a single-piece Al-board with a thickness of 60 mm was fixed to the optical table. It has a length of 2.2 m and a width of 1 m. This board creates the base of the generation chambers. It has threads in a regular pattern which permits fixing optical elements like on a regular optical table. Three vacuum chambers, each with an inner diameter of 500 mm share the breadboard as bottom. The chambers in this part are rubber-sealed and are clamped to the breadboard. The circular shape of the chambers minimizes the required wall thickness of the chambers. A schematic is shown in Fig. 4.7.

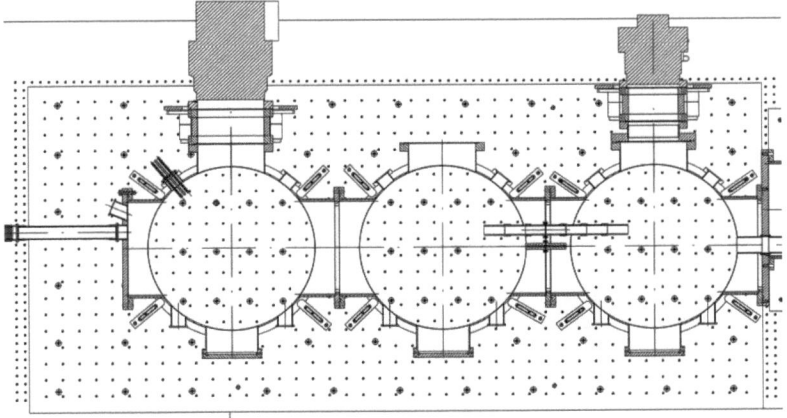

Fig. 4.7 Schematic of the three identical generation chambers on the base plate. The chambers are connected via rectangular flanges. Between the second and third chamber a differential pumping stage is indicated.

To achieve highest stability with the underlying optical table, the board is fixed with screws in a regular pattern to this table. To be able to fix the board also from within the vacuum chambers, sealing screws have been developed. A schematic drawing is shown in Fig. 4.8. Furthermore, the breadboard is temperature stabilized to ±0.1°C which improves the long-term stability of the setup and minimizes mechanical stress on all components, which can be caused by differences in thermal expansion.

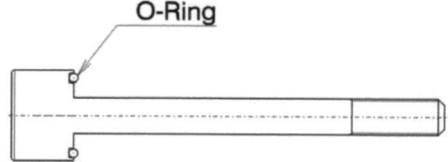

Fig. 4.8 Sealing screw for fixing the aluminum breadboard to the optical table from inside vacuum.

To increase the flexibility in optical layouts the chambers are interconnected by rectangular flanges. They offer a wide aperture in the horizontal plane (a=272 mm) and a small aperture in the vertical plane (b=72 mm) centered at the beam height of the setup. These flanges also serve as holders for metal sheets with holes that enable an easy separation of the chambers to achieve differential pumping. These plates can be exchanged without the need of separating the chambers which allows for easy and fast redesign of the optical beam paths.

A section of the chambers is shown in Fig. 4.9 further illustrating how this part of the beamline is constructed.

The vacuum requirements are not very demanding in this area of the setup. The pressure will be in the range of $1 \cdot 10^{-6}$ mbar when the gas target is not operated, and $5 \cdot 10^{-3}$ mbar when the target is operated. At a pressure of $5 \cdot 10^{-3}$ mbar about 4% of the harmonic radiation is absorbed after one meter of propagation. After the first differential pumping stage, which is approximately 20 cm after the generation target, the pressure is typically $5 \cdot 10^{-5}$ mbar when the target is on, resulting in a negligible absorption of the harmonic radiation. The use of differential pumping stages in an experiment is described in section 5.4.2.

Due to the relaxed vacuum requirements, it was possible to build the lids of the chambers in this section from acrylic glass (thickness 50 mm). They function as large windows and allow for visual inspection of the optical setups.

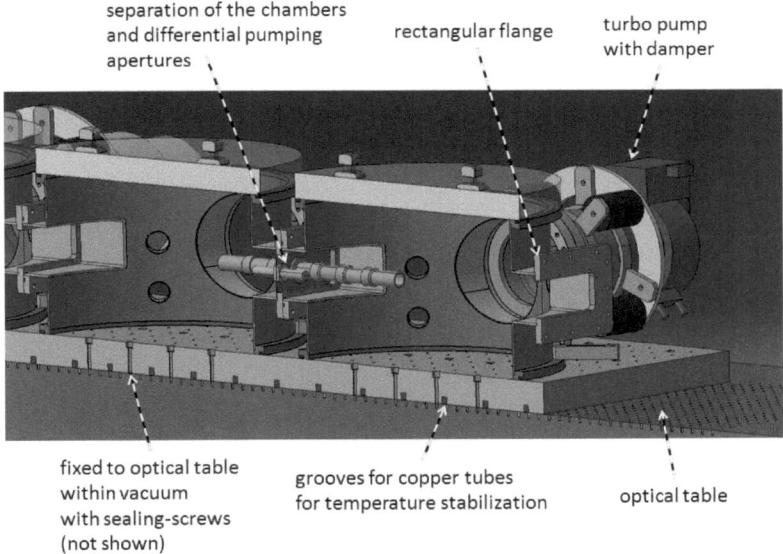

Fig. 4.9 Section of a part of the generation chambers. The holes that allow for fixing the breadboard to the optical table using the sealing screws are visible (Fig. 4.8).

The mechanical stability in this section of the setup allows the implementation of interferometric optical designs with several meters in beam separation. A stability of 65 as (rms) is achieved for the largest type of interferometers as shown in Fig. 4.10. The stability is also proven to be in the attosecond regime by successful experiments with attosecond time resolution (chapter 5 and 6).

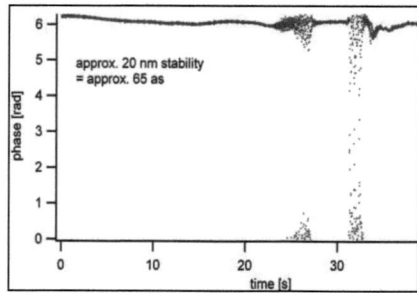

Fig. 4.10 Mechanical stability measured by capturing interferences in an optical interferometer using a Helium Neon laser and a CCD camera (A501k, Basler AG). At two instants (27 s and 32 s) mechanical shocks have been produced on the optical table. The interferometer recovers to its initial position.

4.3.1 Motorized optical components

In this subsection a description of motorized components built into the generation chambers will be given. They enable an in-vacuum operation of optical elements.

- Fine alignment of the optical beams under vacuum is achieved by motorized mirror mounts using picomotors (Model 8203, NewFocus Corporation) or by using motorized optical mounts (Model 8809, NewFocus Corporation).

- Delay stages for interferometrically stable pump probe measurements are built using a combination of a long range translation stage, which is driven by a DC motor (Motor Mike, Oriel Inc.) for rough alignment and a piezoelectric actuator with capacitive feedback for fine positioning (PX100Cap, piezosystem jena GmbH).

- Motorized metallic shutters have been designed. They are operated by a DC motor and allow blocking of a beam, for example the probe beam in a pump-probe setup.

- Irises have been motorized by mounting them on a motorized rotary stage (Model 8401, NewFocus). This enables for energy control and beam shaping. Irises in the laser beam before HHG have a strong influence on the phase matching conditions and are an important tool for optimizing the generation process. We used irises made from stainless steel to have a low absorption of the IR. Because they are used under vacuum anodized material (typically black) would lead to a strong heating of the iris.

4.3.2 Gas targets for high harmonic generation

Several gas targets are available in the setup. A pulsed valve can be used to achieve high density at reduced gas load. A capillary target is available to achieve low density at an extended propagation length. Both targets will be described separately.

4.3.2.1 The pulsed valve

This jet is mounted on a motorized xyz-stage (xy-directions: picomotors Model 8203, NewFocus Corporation; DC-Motor for z-direction = propagation direction of the laser, Oriel Inc). This permits a precise positioning of the target within the laser focus. The travel range in z-direction is 25 mm and allows controlling the phase matching conditions of the high harmonic generation process.

The pulsed valve itself is an improved version of the valve described in [73].

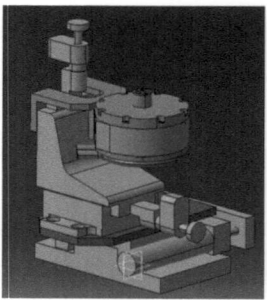 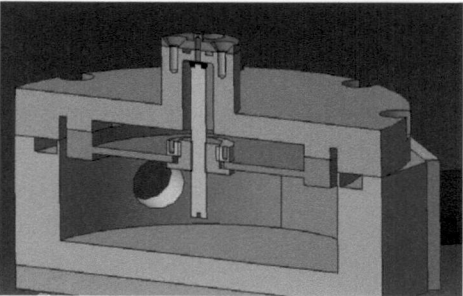

Fig. 4.11 Images of the pulsed valve. On the left the valve is shown with the xyz translation stage. On the right a section of the valve is shown. The red part is a piezoelectric disc which is operated synchronously to the repetition rate of the laser. It moves a pin with an O-ring that seals the valve when no voltage is applied to the piezo.

The gas expands through the exchangeable target tip. Several target geometries have been designed and used in experiments. An open target features a very short interaction length. This relaxes the phase matching requirements, the macroscopic response will be close to the single atom response. This target type was used for achieving the results on QPI (as described in section 3.3).

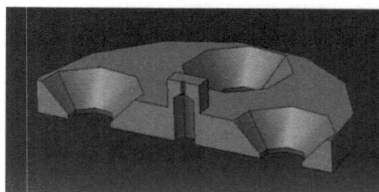 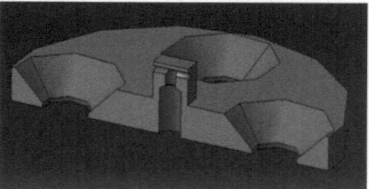

Fig. 4.12 Sections of two target geometries for the pulsed gas jet. The open target (left) offers a short interaction length, the "T"-shaped target (right) offers a high gas density.

Another target type has a "T"-shape and features longer interaction length and higher density [74]. This target is shown in Fig. 4.12 (right). Typical dimensions of the "T" are a length of 1.5 mm and a diameter of 0.3 mm. This target is very useful for the generation of high harmonics in media where a high density is required to obtain a useful harmonic flux. This is for example the case for media with high ionization potential like helium. We have used this target not only to generate harmonics but also as an interaction target (i.e. at the position of the second focus, Fig. 4.2) to significantly absorb high harmonics in a gas. This allows estimating the density in the target to be approximately $5 \cdot 10^{17}$ particles/cm³. With a target length of 1.5 mm this can induce an absorption of approximately 50% of harmonics that are passing through the target (harmonic order 17, absorbing gas helium). It enables the attosecond transient absorption experiment described in chapter 6.

4.3.2.2 The capillary target

A second target is made up by a capillary that is filled from one side with gas and serves as differential pumping aperture. By that a pressure gradient in propagation direction is generated. It is mounted on a translation stage for movement along the propagation direction (DC-Motor, Oriel Inc) on which a motorized Four-Axis Tilt Aligner (NewFocus Corporation) is installed. This enables fine alignment of the capillary in vacuum. This target offers a long interaction length and was used for the generation of harmonics in media with low ionization potential. The gas leak rate is low compared to the pulsed valve. The capillaries that we used usually have an inner diameter of 500 µm. In this case, the capillary functions as a differential pumping aperture only and no guiding effects occur. The capillary target has been used to generate low order harmonics in xenon which have been later used in the attosecond transient absorption experiment described in chapter 6.

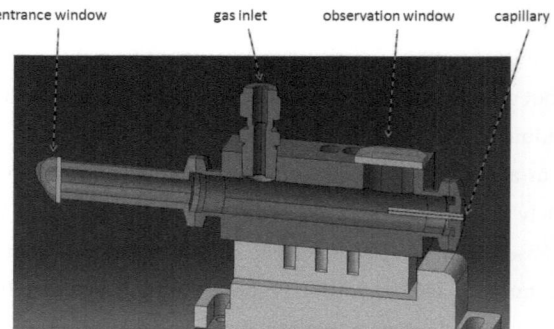

Fig. 4.13 Sectioning of the capillary target. The laser beam enters from left through the entrance window and is focused into the capillary. The capillary forms the gas outlet to vacuum. The enclosed volume is continuously filled with gas for HHG. An observation window allows for visual inspection of the capillary entrance.

4.3.3 Filter wheel for metallic thin-film foils

We designed a new filter wheel for mounting thin-film foils. These metallic foils are used to remove IR radiation and low order harmonics from the high harmonic beam. The wheel offers four mounting position for filters. Each filter frame can be removed separately thus making the exchange of individual filters easy. The wheel is equipped with a protective housing where the filters are stored when the beamline is vented. This prevents the metallic foils from damage. Their thickness is typically in the range of 100 nm which makes them very fragile. The small thickness is necessary to achieve a high transmission in the XUV. In section 5.2 more details on thin-film foils, especially for the case of aluminum will be given. The filter wheel is shown in Fig. 4.14 in an exploded view. The rotation is achieved by a water cooled stepper motor (SP4202, Nanotec GmbH & Co KG). The wheel is directly mounted on the axle of this motor.

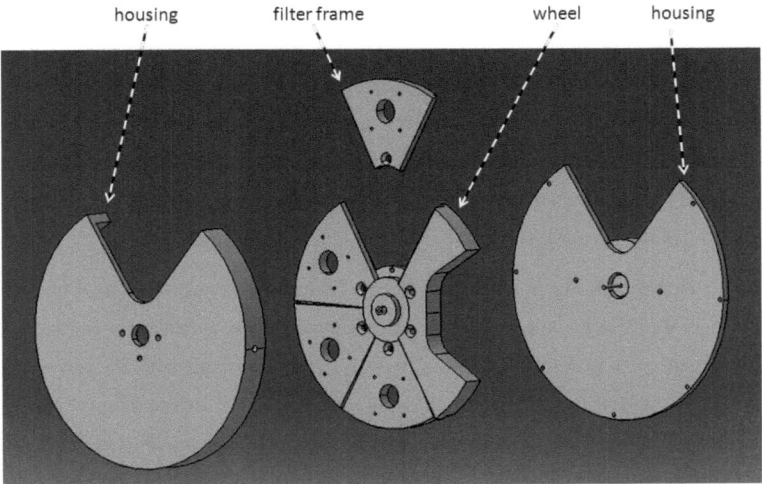

Fig. 4.14 Exploded view of the filter wheel.

4.4 Diagnostics and refocusing chamber

The chamber that follows the generation section has the same size and shape as one of the chambers in the generation part. The lid is made from metal to improve vacuum conditions. In this chamber two instruments for beam diagnostics are installed: A high resolution beam imaging system sensitive in the XUV (Model BIS-1135 with BIS-1-PH probe head with BIS-4135 remote system, Colutron Inc.), as well as a calibrated photodiode (NIST windowless far UV photodiode, NIST National Institute of Standards and Technology, USA) to enable harmonic flux measurements. The toroidal mirror is also installed in this chamber. An image of the chamber is shown in Fig. 4.16.

The beam profiler consists of a microchannel plate (MCP) and a phosphor screen. Incident XUV photons will generate electrons. The number of electrons is amplified by the MCP. The amplification factor can be tuned by the operation voltage of the MCP. Typically it will be in the range of 10^6. The electrons are then accelerated by an electric field to an energy of ~1.5 keV. They hit a phosphor screen which will emit luminescent light. The phosphor is uniformly deposited onto a coherent fiberoptic substrate so that the image can be optically transmitted to outside vacuum of the diagnostics chamber. For image readout a CCD camera chip is in direct contact to the end of the fiber optic. For the operation of a MCP a pressure in the range of $5 \cdot 10^{-6}$ mbar is required. This is achieved even under operation of the gas targets in the generation chamber by the differential pumping apertures that reduce the leak rate between the individual chambers (see also section 5.4.2).

The XUV photodiode is made up of two elements: A stainless steel cylinder which is the anode and the photo cathode, which is formed by high-purity aluminum on a polished quartz substrate. It is schematically shown in Fig. 4.15. XUV radiation hitting the cathode will cause the emission of electrons. The assisting field created by the anode, which is operated at a voltage of +60 V, leads to a current. A calibrated electrometer is used to obtain absolute flux data (Model 6514/E, Keithley Instruments Inc.). The quantum efficiency (electrons/photon) of the diode is in the range of 0.1 to 0.15.

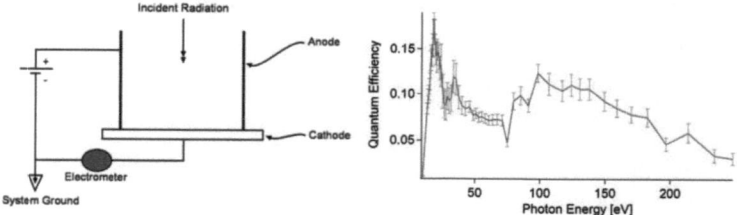

Fig. 4.15 Schematic of the calibrated photodiode from NIST (National Institute of Standards and Technology, USA) and its quantum efficiency (electrons per photon).

Both devices are mounted on a translation stage that is driven by an arbor which is rotated by a water cooled stepper motor (ST4209, Nanotec GmbH & Co KG).

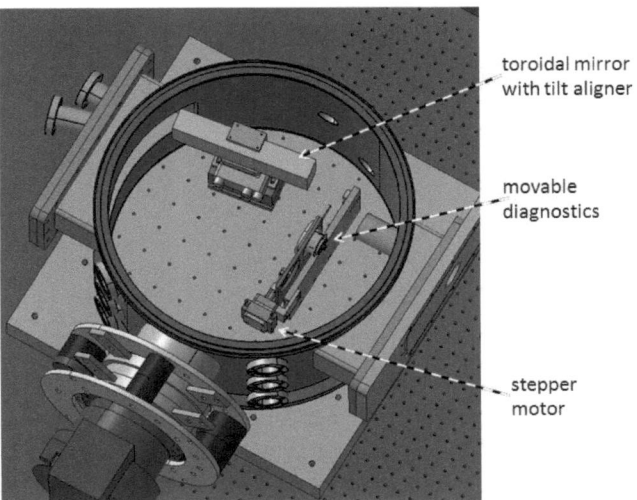

Fig. 4.16 View into the diagnostics and refocusing chamber. The beam enters from the right.

The toroidal mirror is installed in this chamber. It images the harmonic generation focal spot to a spot in the interaction region. This chamber is only clamped to the optical table. It has not a fixed position, but can be moved in case the generation focal spot has to be at another position for a specific experiment. Typically we have the generation target located at the end of the second generation chamber as shown in Fig. 4.2. For the alignment of the toroidal mirror a motorized Five-Axis Tilt Aligner (NewFocus Inc.) is being used. It permits movement in all directions required for the alignment of a toroidal mirror. A rotation of the mirror around its meridial axis is only indirectly possible. Due to the toroidal symmetry, a rotation is identical to a change of the total height of the mirror. Since the alignment of the toroidal mirror is very critical, a way to characterize the created focal spot by further diagnostics is required. This measurement has to be possible while the setup is under vacuum. It is achieved by installing a small cross chamber following the focusing chamber (see Fig. 4.6). A two-inch mirror mounted on a push pull translator (Linear motion push pull translator, Caburn-MDC Europe Limited) can be inserted to send the IR beam through an optical window out of vacuum. It is sent to an IR beam profiler (Beamage-CCD12, Gentec EO) that is installed on a translation stage. It allows for scanning and measuring the focal spot created by the toroidal mirror over more than its Rayleigh-range. The cross chamber also serves as additional differential pumping stage, reducing the gas flow from the generation section into the interaction chamber.

4.5 Interaction chamber

An electron time-of-flight (TOF) spectrometer was designed and installed in the interaction region and was used for attosecond experiments. The spectrometer is field-free. A mu-metal [75] tube is installed inside the TOF-tube to shield it from magnetic fields (Fig. 4.17). The entrance of the tube is made from pure mu-metal. It has entrance/exit holes for the laser beam and an opening for inserting a diffuse gas target (Fig. 4.18). The target is made of an aluminum needle that can be moved close to the interaction focal spot by an xyz-manipulator which can be operated from outside vacuum manually (EC Manipulator – EC-1.39-2, Hositrad vacuum products).

The free electrons that will be generated in an experiment travel a distance of 1 meter inside the mu metal tube, then pass a grid on ground level and are then accelerated onto a high speed MCP time-of-flight detector (F4655-12X, Hamamatsu Photonics). The energy of the electrons is thereby increased from the eV regime to the keV regime where the MCP can detect the electrons. The detector has an active area of 14.5 mm in diameter. This geometry leads to an collection angle of 0.4 deg. The typical rise time of the MCP signal is 400 ps and a typical pulse width is 750 ps (FWHM). To increase the collection angle to 1.8 degrees and thus the electron count rate a short flight tube (23 cm) is available. It has a decreased energy resolution. The TOF signals are analyzed by a time-of-flight analyzer (P7889, FAST ComTec GmbH). This acquisition card offers multiple event time digitizing with a resolution of 100 ps.

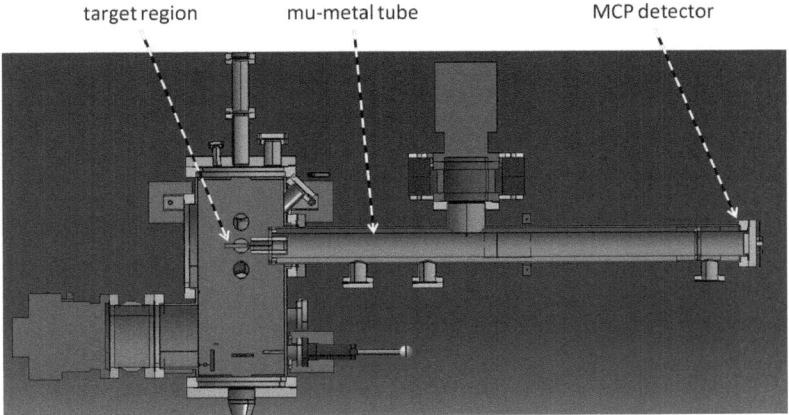

Fig. 4.17 Section of the interaction chamber with the long time-of-flight tube attached. The laser beam enters from the bottom. The TOF tube is made from stainless steel with a 1 mm thick mu-metal tube inside to prevent magnetic fields from influencing the propagation of the electrons.

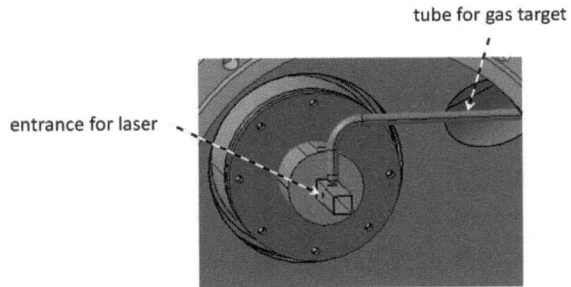

Fig. 4.18 The entrance of the time-of-flight tube. The gas target is shown in red. It is moved into the mu-metal shielding of the TOF-tube.

The resolution of the TOF spectrometer can be calculated as follows:

Let s be the length of the TOF tube, v_0 the initial velocity of the first electron. A second electron has a velocity of $v_0 + \Delta v$. The time-of-flight will then be $t_0 = s/v_0$ for the first electron and $t_1 = s/(v_0 + \Delta v)$ for the second electron. The difference between these two times has to be larger than the resolution of the detector. This equation results in a resolvable difference in velocity of

$$\Delta v = \frac{tv_0^2}{s - tv_0}.$$

We then get the resolution by $\Delta E / E_0 = \Delta v / v_0$. We will calculate some typical values (based on the photon energy of the Ti:Sapphire laser, 1.5 eV). For most attosecond experiments a resolution of $\Delta E < 1.5$ eV will be required.

- For the case of low order harmonics (harmonic order 15) ionizing argon (I_P=15.75 eV) the photoelectrons will have an energy of approximately 7 eV. Using the short TOF tube, the resolution will be $\Delta E = 1.9 \cdot 10^{-2}$ eV and $\Delta E / E = 2.7 \cdot 10^{-3}$.

- For a high harmonic order (harmonic order 91) ionizing argon, the resolution will be $\Delta E = 1.4$ eV in case of the short tube, which is not good enough. With the long tube the resolution is improved to $\Delta E = 0.3$ eV.

4.6 Refocusing and XUV spectrometer

After passing the interaction chamber the radiation is focused a second time. A spherical mirror under grazing incidence is used in this case as described in section 4.1. The dimensions of the generated focus are well suited for the entrance slit of the XUV spectrometer which has a width of 100 μm. Since we only measure spectra, the spatial and temporal properties of the beam are less important, which relaxes the requirements regarding the surface quality of the mirror. The mirror is mounted in a standard CF160 cross chamber and movable motorized by using rotation stages and DC motors (Motor Mike, Oriel Inc.).

The XUV spectrometer (model 251, McPherson Inc.) is equipped with a grating (450 grooves/mm) that has a calculated first order efficiency of 4 to 7 percent. It is platinum coated and is used at an angle of incidence of 70.5 degrees. The diffracted radiation is finally detected using a CCD (PIXIS-XO, 400B, Princeton Instruments). It has 1340 x 400 imaging pixels, offers a quantum efficiency of 20 to 50 percent and is capable of single shot measurements at 1 kHz repetition rate. This camera is sensitive in the IR.

Therefore a filter holder for thin-film filter foils is installed before the spectrometer to remove IR radiation from the beam. Two filter frames are available on a push pull translator (Linear motion push pull translator, Caburn-MDC Europe Limited). In combination with the filter wheel in the generation chamber this gives the possibility of characterizing the transmission of the filters as discussed in section 5.2.

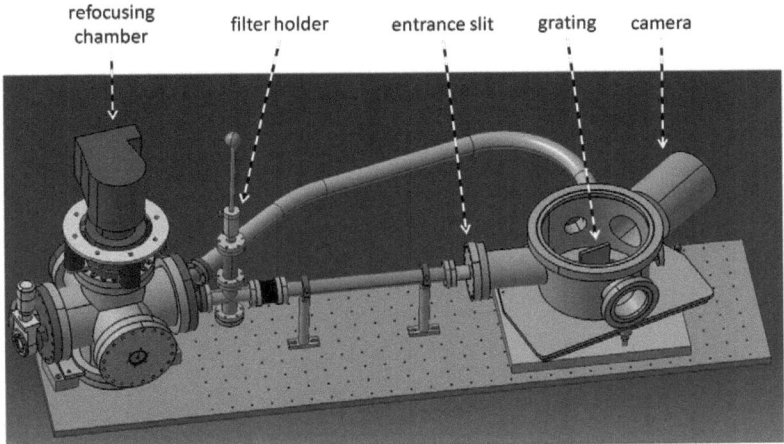

Fig. 4.19 Spectrometer unit with refocusing chamber and the commercial XUV spectrometer.

The spectrometer unit (refocusing and spectrometer chamber) is mounted on an aluminum board and can be moved on the optical table. This allows quick redesign of the beamline if necessary. The whole unit is pumped by a single turbomolecular pump mounted on the cross. Since no MCP is operated in the spectrometer itself, it can be pumped indirectly via a flexible tube that connects the spectrometer to the cross.

4.7 Active beam stabilization

The laser systems used for the experiments are installed in another laboratory than the beamline. The laser beam is sent through a long vacuum tube (length

~6 m) directly into the harmonic beam line. To enable measurements with a long acquisition time it is necessary to stabilize the beam pointing of the laser into the beamline. A glass plate (fused silica) is mounted under Brewster angle in a small chamber in front of the harmonic beamline (see Fig. 4.6). The reflected light is intense enough to be detected on a CCD camera (Webcam, Logitech Quickcam IM/connect, Logitec Inc.). The center of gravity of the measured beam profile is calculated and a feedback loop to a mirror mounted on a piezo tip / tilt stage (piezosystem jena GmbH) is employed. This enabled measurements with attosecond resolution with an acquisition time of more than 30 hours.

Chapter 5

Characterization of APTs

In this chapter, measurements that proof the functionality of the high harmonic beamline are shown. These experiments are as follows: A beam profile measurement of high harmonic radiation, a measurement of the transmission of an aluminum thin-film filter and a flux measurement of the high harmonics. Besides the measurement of the filter transmission further properties of thin-film filter foils are discussed in section 5.2. In section 5.4 an experiment with attosecond resolution is presented: The characterization of attosecond pulse trains generated in argon and xenon using the RABBITT technique [25].

5.1 A beam profile of high harmonic radiation

In Fig. 5.1 a typical beam profile with an according line profile (integration along a narrow line through the center of the beam) of the high-order harmonics is shown. In this case the harmonics have been generated in argon using the pulsed "T"-target (Fig. 4.11). The focus was before the jet where the short trajectory is phase-matched preferably.

A Gaussian fit of the line profile agrees well with the measurement. We found that in most cases the beam profile of the harmonics is close to Gaussian. Imperfections in the generating laser beam typically won't be focused well. Due to the high nonlinearity of the generation process, these parts don't generate high harmonics. However, the beam profile can be distorted if the intensity is too high or by a non-uniform transmission of a thin-film filter foil (see Sec. 5.2).

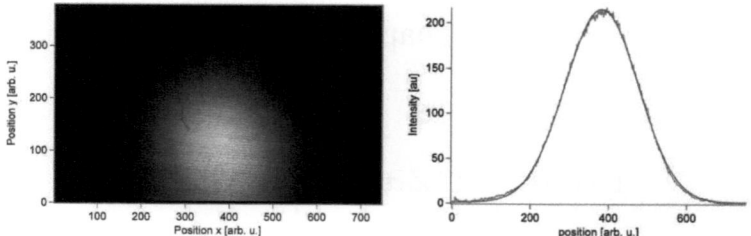

Fig. 5.1 Typical beam profile of high harmonics. On the left the CCD image is shown. The right graph shows an according line profile (red) with a Gaussian fit (blue).

5.2 Properties of thin-film filters, measurement of the transmission of an aluminum filter foil

In Fig. 5.2 the calculated transmissions of an aluminum and a zirconium foil is plotted [71]. The thickness of the filters is 100 nm. The transmission in the visible is in the range of $5 \cdot 10^{-8}$ for the aluminum filter and $5 \cdot 10^{-6}$ for the zirconium filter [76].

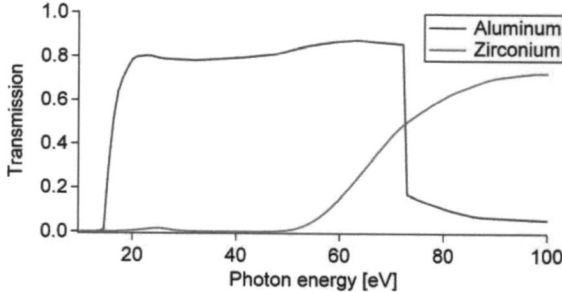

Fig. 5.2 Transmission of thin-film foils for the case of aluminum and zirconium, each with a thickness of 100 nm [71].

We mostly used thin-film foils of aluminum in our experiments. An oxide layer on an aluminum filter will dramatically reduce the transmission [77]. To reduce oxidation during transportation of the filters we manufactured aluminum filters within the physics department at ETH. A thin layer of victawet (Structure Probe, Inc.) is evaporated on a substrate. In the next step the aluminum is evaporated and an aluminum layer with the required thickness is formed on the victawet layer. Victawet is a wetting agent, by moving the substrate into a water bath, the aluminum foil will be detached and will float on the water surface. It can be removed using a filter frame, thereby forming a free standing filter foil. The step of shipping a filter is thereby eliminated and the thickness of an initial oxide layer is reduced. Also it is fast to get new filters with different thicknesses.

The beamline offers the possibility to use two thin-film filter foils simultaneously. This can be used to characterize the transmission of the filters. The first filter with a transmission $t_1(\omega)$ is installed in the filter wheel (Fig. 4.14) in the generation section of the beamline, the second filter with a transmission $t_2(\omega)$ is installed before the XUV spectrometer (Fig. 4.19). High harmonic radiation is generated and detected in the spectrometer. Assuming that the measured spectrum without any filter is $A(\omega)$, the measured spectrum with both filters inserted is $A_{12}(\omega) = t_1(\omega) \cdot t_2(\omega) \cdot A(\omega)$, whereas the spectrum with the first filter only is given by $A_1(\omega) = t_1(\omega) \cdot A(\omega)$. The transmission of the second filter can then be calculated to be

$$t_2(\omega) = \frac{A_{12}(\omega)}{A_1(\omega)}.$$

In Fig. 5.3 the measured transmission is shown for an aluminum filter with a thickness of 120 nm. The spectra $A_{12}(\omega)$ and $A_1(\omega)$ are shown. The harmonics have been generated in argon. Data points for calculating the transmission are taken at the peaks of the individual harmonics. The measured transmission is shown in blue in the right graph. It is worse than the expected transmission of an aluminum filter with a thickness of 120 nm [71]. In addition, the slope of the measured transmission is much steeper than the calculated transmission, which is almost flat. This behavior is a result of an oxide layer. The estimated thickness of oxide is approximately 2 nm (see Fig. 21 in [77]).

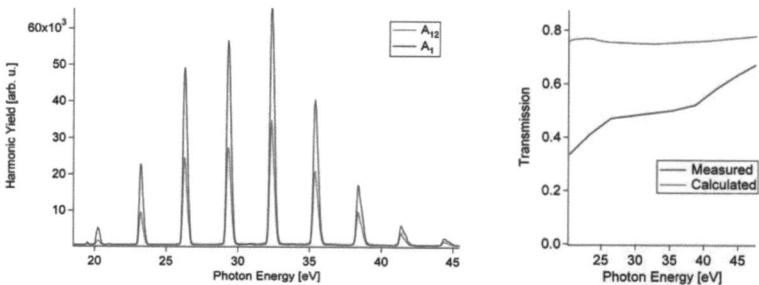

Fig. 5.3 Left: Two harmonic spectra generated in argon with identical conditions. One spectrum (A_1) is recorded with one filter (shown in blue), the second spectrum (shown in red) is recorded with two filters (A_{12}). This allows characterizing the transmission of the filter. **Right:** The measured transmission t_2 is shown in blue and the calculated transmission for unoxidized aluminum is shown in red.

In addition to the spectral transmission we can characterize the spatial transmission profile of the thin-film filter foils by performing high harmonic beam profile measurements. In Fig. 5.4 two profiles with spatial distortions are shown. These imperfections cannot be found by visual inspection of the filters. We found that performing an experiment with attosecond resolution can be very difficult with such a defective filter. Usually these defects are localized on a small area on the filter and by changing the position on the filter one can avoid them.

Fig. 5.4 Beam profiles of the high harmonics transmitted through a thin-film foil of aluminum with imperfections. The profile in the left image shows some droplet-like structures, the filter in the right image shows a kink. These imperfections might have been created during manufacturing of the filters.

In addition to the transmission of these filters, their dispersion can be an interesting property. For example, the group velocity dispersion of an aluminum foil compensates for the intrinsic harmonic chirp of the short electron quantum path in a spectral range from 20 eV to 55 eV. Thereby an attosecond emission from the short trajectory can be compressed in time [78]. For that reason the group velocity in the typical spectral range of high harmonics is shown in Fig. 5.5 for aluminum. The refractive index data have been taken from [79].

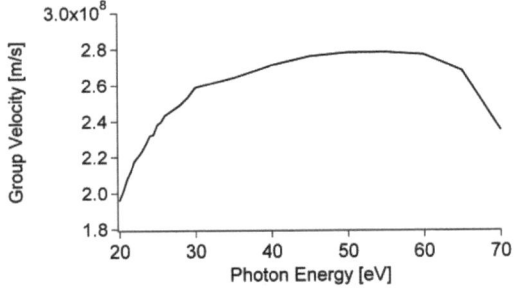

Fig. 5.5 Calculated group velocity versus photon energy for aluminum. Data of the refractive index has been taken from [79].

5.3 Measurement of the flux of high harmonics

The photodiode (Fig. 4.15) can be used to calibrate a high harmonic spectrum to absolute photon numbers or energy. In the following a complete description of the calibration procedure is given. For the calibration a high harmonic spectrum has to be acquired with the XUV spectrometer. Let this spectrum be $S(\omega)$. In addition, the absolute number of generated electrons per shot $N_{cal.}$ in the calibrated photodiode has to be measured. To calibrate the spectrum one has to know the spectral shape at the position of the photodiode. To calculate this spectrum, the spectral transmission of all additional optical elements from the position of the diode to the spectrometer has to be known. In the experimental layout of our beamline these optical components are two gold mirrors and the grating in the spectrometer, not to forget the spectral response of the camera of the spectrometer. Let the total spectral transmission of all these components be $T(\omega)$. For most elements one has to use calculated values as they can be obtained from [71] or [72] and data from technical datasheets. These values might not be known precisely and therefore introduce a large error in the measurement.

The spectral shape at the photodiode is given by $S_{diode}(\omega) = S(\omega)/T(\omega)$. By weighting this spectrum with the quantum efficiency of the diode $S_{electrons}(\omega) = Q_{eff}(\omega) \cdot S_{diode}(\omega)$ (see Fig. 4.15), photon flux will be converted to electron flux ($[Q_{eff}(\omega)]$ = electrons per photon). To achieve the absolute calibration, a spectral integration has to be performed. The result gives the total number of electrons $N_{uncal.} = \int S_{electrons}(\omega)d\omega$. This quantity has been measured with the electrometer and the photodiode $N_{cal.}$. Thus the conversion factor f for an absolute calibration is $f = N_{cal.}/N_{uncal.}$. This factor can then also be used for the absolute calibration of the flux axis of the spectrum $S_{calibrated}(\omega) = S_{diode}(\omega) \cdot f$.

If one wishes to calculate the spectrum at the source, one has to take the aluminum filter in the generation part of the setup into account. This filter has to be inserted during the measurement and is present in the measurement of the spectrum and of the flux. It can be characterized with good precision as described in section 5.2.

In the example measurement we generated high harmonics in argon using the "T"-shaped target. The calibrated spectrum is shown in Fig. 5.6 (at the position of the source), a total energy of 1.4 nJ per shot was estimated. The generation was done with IR pulses with an energy of 450 µJ, which results in a total energy conversion efficiency of $3.1 \cdot 10^{-6}$.

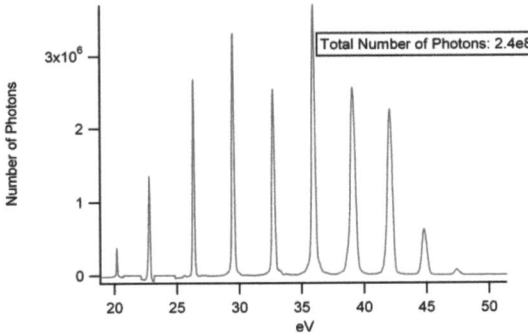

Fig. 5.6 Calibrated high harmonic spectrum generated in argon. The total harmonic energy content per laser shot is 1.4 nJ. The conversion efficiency is $3.1 \cdot 10^{-6}$.

5.4 Temporal characterization of an APT

The superposition of several plateau harmonics from a single type of electron trajectory (e.g. the short trajectory) generated by a multicycle driving pulse leads to the generation of an attosecond pulse train. The selection of the trajectory can be achieved by proper phase-matching conditions and/or by spatial filtering (Sec. 2.1.4). Assuming, in first approximation, a constant phase difference φ between consecutive harmonics and constant harmonic amplitude, the intensity generated by N harmonics is given by

$$I(t) \propto \left| \sum_{q=q_0}^{q_0+N-1} e^{-i(2q+1)\varphi} e^{-i(2q+1)\omega t} \right|^2 .$$

This describes a train of pulses in time domain with a periodicity of half the laser period $T_{Laser}/2 = \pi/\omega$ as expected from the three step model. q_0 is the lowest harmonic order taken into account. The duration of a single pulse is in the range of $\tau \approx T_L/2N$ (see also Sec. 2.1.5).

In reality the phase-differences between consecutive harmonics will not be constant but will change with harmonic order. This harmonic chirp will influence the pulse duration of the individual attosecond pulses in the APT, it is therefore referred to as "atto-chirp". The goal of a RABBITT measurement is to determine these phase differences [25].

5.4.1 Theoretical background of RABBITT

Pulses in an attosecond pulse train don't typically have enough energy to induce nonlinear effects in a target atomic medium. Each harmonic will therefore ionize the atom in a single-photon transition. However, the combination of an attosecond pulse train with a strong infrared field can lead to a multiphoton transition in the continuum [80]. Thereby the absorption of a harmonic photon can go along with either an absorption or emission of an IR photon. The interaction can be restricted to these two photon transitions if the infrared intensity is not too high (approximately $5 \cdot 10^{11}$ W/cm^2). It will result in the generation of sidebands in the electron spectrum as is illustrated in Fig. 5.7.

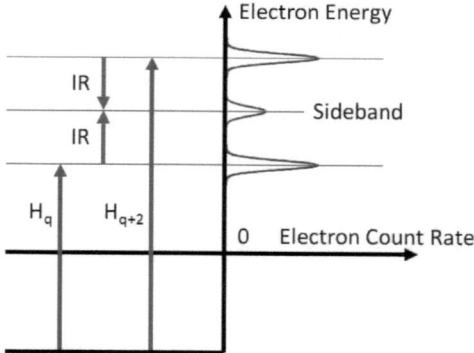

Fig. 5.7 Principle of the RABBITT technique. A sideband signal is generated in an electron time-of-flight spectrum. It is a result from interference between the absorption of a high harmonic photon with order q and an infrared photon (IR) and the absorption of a high harmonic photon with order q+2 and the emission of an infrared photon (IR). The sideband amplitude is a function of the delay between the XUV pump and IR probe pulse.

Since only odd harmonics are contained in the harmonic spectrum generated by a multicycle driving field, the energy of a sideband electron will correspond to an even harmonic. The amplitude of the sideband signal S depends on several factors. It is derived by a quantum mechanical calculation in [25] and more detailed in [81]. It results in

$$S_{q+1}(\omega,\tau) \propto \cos(2\varphi_{IR} + \Phi_q - \Phi_{q+2} + \Delta\varphi_{atomic}^{q+1}).$$

The sideband signal depends on the phase difference between consecutive harmonics $\Delta\Phi_{q+1} = \Phi_q - \Phi_{q+2}$, on an atomic phase $\Delta\varphi_{atomic}^{q+1}$ and also on the phase of the IR φ_{IR}. The atomic phase has to be obtained from theoretical calculations. Typically the contribution of the atomic phase is small. It depends on the sideband number and on the atom that is used for the measurement (a table with numerical values can be found in [25]). The phase of the IR field changes with time delay τ between XUV and IR: $\varphi_{IR} = \omega_{IR}\tau$.

Thus, the sideband amplitude can be controlled by the strong (compared to the XUV) infrared field. By recording electron time-of-flight spectra for different delays τ and extracting the sideband amplitudes as a function of delay, an oscillating signal with half-cycle periodicity will be retrieved. The phase differences $\Delta\Phi_{q+1}$ can be determined by means of a Fourier analysis for each sideband.

Finally by concatenation of the phase differences $\Delta\Phi_{q+1}$, the relative phases Φ_q of all harmonics can be calculated. The phase of the lowest order can be assumed to be zero, a constant offset would only change the position of the electric field under the carrier of the attosecond pulses. The variation of the phase of the individual harmonics, the "atto-chirp", will influence the attosecond structure in the APT.

Now that the phases of the harmonics are found, a spectrum has to be obtained for reconstructing the attosecond pulse. If possible, this should be done using the XUV spectrometer, which offers a better signal to noise contrast compared to the harmonic spectra measured in the TOF spectrometer. Thereby the harmonic intensities I_q are measured. The reconstructed average pulse in the train of pulses is then given by the following Fourier series

$$I(t) = \left| \sum_q \sqrt{I_q} e^{-iq\omega t - i\Phi_q} \right|^2.$$

In that case, the individual harmonic amplitudes are assumed to be monochromatic. The corresponding APT would contain an infinite number of attosecond pulses, each separated by half a fundamental laser cycle. Thus the finite duration of the driving field and any variations of the attosecond pulse characteristics over the pulse train are neglected. These variations can be caused by a change in intensity in the generation of the APT (caused by the IR pulse envelope) or by a chirp in the generating IR pulse. The properties of the APT on a femtosecond timescale, which is caused by a phase variation over a single harmonic (called the "femto-chirp"), cannot be determined by RABBITT. This also means that the number of pulses in the train cannot be characterized by this technique.

5.4.2 Experimental setup

The acquisition of a RABBITT trace is achieved in a typical XUV/IR pump-probe setup as indicated in Fig. 4.1. We implemented it as a Mach-Zehnder like interferometer. In this scheme, sub-half-cycle stability is required, which can be reached in the new high harmonic beamline.

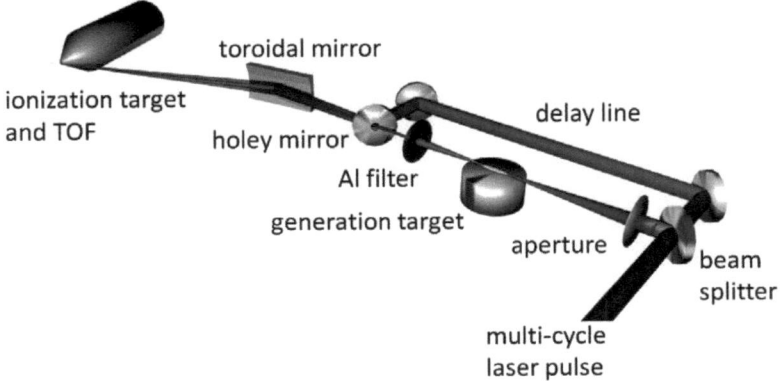

Fig. 5.8 Schematic of the experimental setup for the RABBITT measurement. A detailed description can be found in the text.

The laser pulses are split into two beams using a beam splitter. These form the two arms of the interferometer. In the first arm the beam passes an aperture and is focused into the generation target where the APT will be produced. The aperture is used to optimize the phase-matching conditions in the generation process. The IR radiation is removed using a thin foil of aluminum. The second arm propagates via a controllable delay stage. The two beams (XUV from the first arm and IR from the second) are combined using a mirror with a center hole. No transparent optics are available in the XUV spectral region. Therefore this holey mirror is the only possibility to recombine the two beams collinearly. An attosecond measurement cannot be done with crossed beams, because of the resulting time-smearing of the crossed wave-fronts. By the holey mirror, a beam with XUV in the center surrounded by a delayable IR field is generated. The two beams are focused into the interaction target of the electron time-of-flight spectrometer and the generated photoelectrons are detected.

In the generation part of the setup (Fig. 4.7), the pump-probe scheme was implemented as shown in Fig. 5.9. The laser pulses enter from the left and are split into two beams by the beam-splitter. One beam is focused by a spherical mirror (FOC.) into the target (here, the capillary target is shown). Before focusing, the beam is passing a motorized iris and a retro-reflector installed on a piezo-controlled delay-stage. After generation, the beam passes the filterwheel (Fig. 4.14) and the holey mirror. In this case, a flat silver mirror with an outer diameter of 25 mm was used. The hole diameter was 5.5 mm (Article number 104764, Layertec GmbH).

The second beam passes a motorized iris which allows for an adjustment of the IR intensity in the interaction region. Its divergence is matched to the divergence of the XUV by a defocusing mirror (DEF) before it recombines with the XUV on the mirror with the center hole. This is important in order to achieve the same focusing behavior for both the XUV and IR with the toroidal mirror. A mismatch in the wavefronts of the two beams will result in a time-smearing in the pump-probe measurement which can make the RABBITT measurement impossible.

Let l be the distance from the high harmonic generation target to the holey mirror and a the distance from the holey mirror to the defocusing mirror (DEF). In that case, the focal length f of the defocusing optics has to be $f = l - a$ for a match in divergence.

In Fig. 5.9, the use of differential pumping apertures is illustrated. A long tube with an inner diameter of 30 mm is used to transfer the beam from the left chamber to the right chamber. For the XUV beam a short tube with a small diameter can be used because this beam is focused and small. The resistance Z for a molecular flow of a long tube $x \gg y$ (length x, diameter y) is given by

$$Z = \frac{12 \cdot x}{y^3 \cdot \pi \cdot \overline{c}} \propto \frac{x}{y^3}.$$

Thus, a larger diameter cannot be easily compensated for by increasing the length of the tube. Here $\overline{c}$ is the average velocity of the gas in the chambers [82].

In our experiment we improve the pressure from $5 \cdot 10^{-3}$ mbar in the left chamber to $5 \cdot 10^{-5}$ mbar in the right chamber. The pressure in the following chamber (characterization and focusing) is further improved by a second differential pumping stage and permits for the operation of the microchannel plate (MCP) of the high harmonic beam-profiler which requires a pressure in the 10^{-6} mbar range or less.

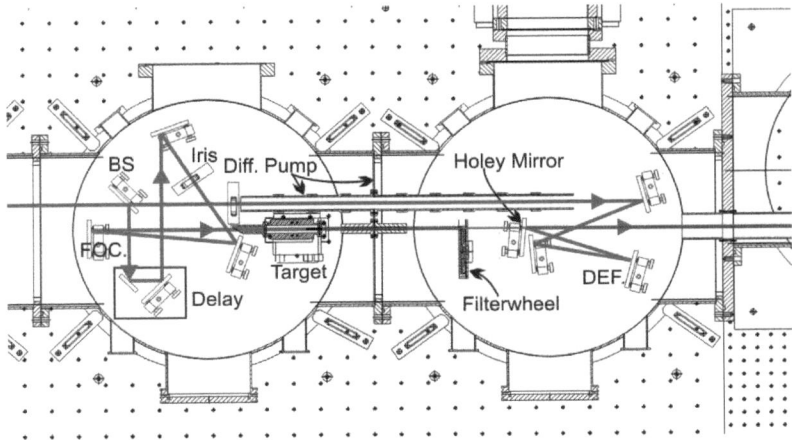

Fig. 5.9 Experimental realization of the optical setup for a RABBITT measurement in the high harmonic beamline. A detailed description can be found in the text. BS=beam splitter.

The combined beam is focused into the interaction region by the toroidal mirror. The atoms will be ionized and the generated photoelectrons are detected in the TOF spectrometer. For good experimental results perfect temporal and spatial overlap between the beams is required. Furthermore the toroidal mirror has to be very well aligned. The alignment can be done by removing the thin-film filter and detecting the combined IR/IR beam on a CCD and on a IR spectrometer. This can be done by inserting the outcoupling mirror that is installed after the focusing chamber (see section 4.4 and Fig. 4.6). In the first step, the beam profile of the IR beams can be checked. A typical profile of the IR probe beam has a Gaussian shape and a measured diameter of 60 μm to 80 μm in the focal spot. This beam has to be overlapped with the second beam.

Spectral interferometry can be used to find the temporal overlap between the two infrared pulses. Fine alignment of the optical paths has to be done when the setup is evacuated, because the strong forces caused by the pressure difference will result in minor misalignment after pumping of the setup. To achieve this, two mirror mounts in the generation arm and two in the probe arm are motorized. This gives full control of the alignment of the interferometer. When overlap in time and space is achieved, the delay between the pulses can be changed using the delay stage. Observing the resulting interference structure between the two pulses on the CCD beam profiler allows for a final evaluation of the alignment of the interferometer as shown in Fig. 5.10.

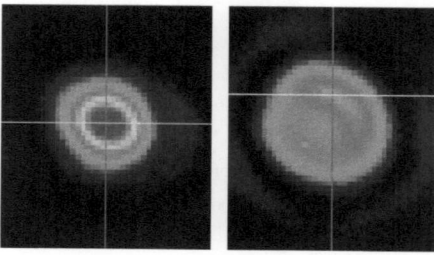

Fig. 5.10 Beam profile in the interaction region. Both IR beams are present and overlap in time and space. The two images show the situation for two different time delays: constructive (left) and destructive (right) interference. The situation shown here is good enough to allow for an attosecond time-resolved measurement.

5.4.3 Experimental results

RABBITT traces have been recorded for high harmonics generated in argon and xenon. The photoelectrons have been generated in argon gas in both cases. In this section the results of the attosecond reconstruction will be presented. The attosecond pulse train generated in xenon will be used in the attosecond transient absorption experiment described in chapter 6.

For the case of argon as generation medium, the harmonic emission that has been calibrated to absolute photon numbers (Fig. 5.6) has been temporally characterized. The reconstruction yields an average pulse duration of 450 as (in the interaction region). The Fourier-transform limit is 160 as. The measurement has been done with an aluminum filter thickness of 120 nm after the generation. The energy of the probe pulses was 5 μJ. The attosecond pulses are far longer than their Fourier limit which is due to the intrinsic harmonic chirp. In Fig. 5.11 the harmonic spectrum with the relative delays is shown. The relative phases show a almost linear behavior which results in a strong quadratic phase behavior in the phases of the harmonics.

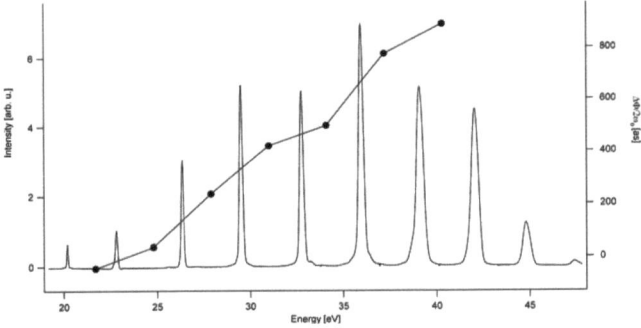

Fig. 5.11 Harmonic spectrum (left axis) generated in argon with their corresponding relative delay, line as guide to the eye (right axis).

Pulses generated from the short electron quantum path possess a positive chirp (i.e. the lower energetic photons are emitted before the higher energetic photons, see Fig. 2.5). This chirp can be compensated for by making use of the material properties of aluminum [79, 83]. The dispersion of atoms is negative after an absorption edge (the group velocity for photons with higher energy is larger than for lower energetic photons, see Fig. 5.5). This method has been investigated and demonstrated experimentally in [78]. In our case we estimate that the chirp can be compensated for by an aluminum foil with a total thickness of approximately 500 nm.

For the generation of high-order harmonics in xenon the capillary target shown in Fig. 4.13 has been used. The capillary diameter used was 500 μm and the capillary length was 15 mm. The focus was positioned at the exit of the capillary and at a pressure of 24 mbar (at the capillary inlet) the high harmonic generation was optimized in yield. The RABBITT trace was generated in argon and is shown in Fig. 5.12. The color scale was changed in this figure at a delay of ~ 4 fs for better visibility of the different harmonic orders. The number of harmonics generated in xenon is very low compared to the case of argon. The stepsize in delay was 107 as, data has been acquired over 60000 lasershots per step. The aluminum filter used in this measurement had a thickness of 100 nm. Clear oszillations of the sideband amplitudes are visible. This proofs that an APT was generated in xenon and allowed to reconstruct an average pulse duration of 400 as. The harmonic spectrum with the relative delays is shown in Fig. 5.13. The reconstruction in this case is not very trustfull because the number of harmonics and thus the number of sidebands is very low.

However, the measurement is very important since it proofs the existence of an APT. This APT will be used in an attosecond experiment presented in Chapter 6.

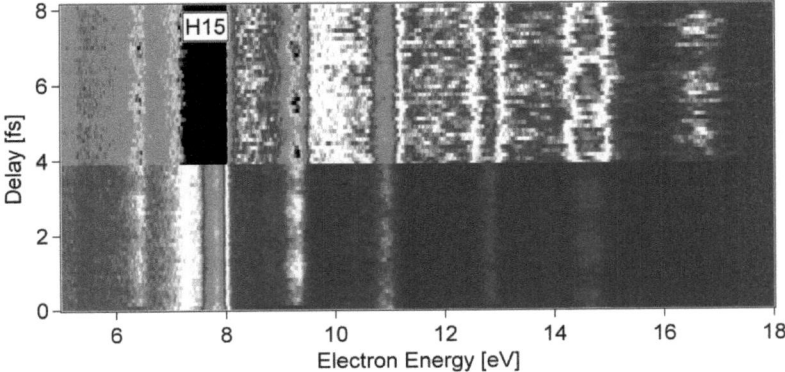

Fig. 5.12 RABBIT trace for a high harmonic generation in xenon. The photoelectrons have been generated in argon. The color scale was changed at a delay of ~ 4 fs for better visibility.

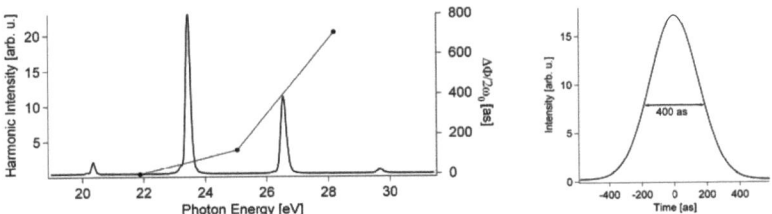

Fig. 5.13 (left) High harmonic spectrum (left axis) generated in xenon with the relative delays (right axis). The corrsponding RABBITT trace is shown in Fig. 5.12; **(right)** Reconstructed average pulse.

Chapter 6

Attosecond Transient Absorption

The new beamline (see Chapter 4) allows for experiments with attosecond time resolution using attosecond pulses from high harmonic generation. Its functionality has been proven in a RABBITT measurement where an attosecond pulse train generated in xenon was temporally characterized (see Sec. 5.4). We will use this APT to perform a new type of attosecond experiment: We spectrally resolve the absorption of the APT in a helium target. We can control the absorption behavior with a strong infrared field with attosecond resolution.

Typically, attosecond experiments use the detection of photoelectrons or –ions, like the RABBITT measurement. In contrast, in the experiment presented here, the transmitted photon yield is measured with the XUV spectrometer. We find that the photon yield not only modulates on the timescale of the envelope of the IR pulse, but also shows oscillations with half-cycle periodicity of the fundamental laser field. In addition, a strong asymmetric behavior of the absorption close to the ionization threshold with a localized enhancement was found. This experiment is a first step towards all-optical experiments in the attosecond domain, which offer the advantage of fast data acquisition and excellent signal to noise ratio.

6.1 Experimental setup

The experimental setup for the attosecond transient absorption is very similar to the one that has been used for the RABBITT measurements. Laser pulses with a duration of 30 fs and a center wavelength of 800 nm are split in two beams using a beam splitter. One beam is used to generate harmonics in xenon. We confirmed that the obtained harmonics (orders 13 to 19) form an

APT with an average pulse duration of ~400 as (see Sec. 5.4.3). The IR radiation is removed from the harmonics using an aluminum foil with a thickness of either 100 nm or 500 nm. The harmonics are recombined with the second beam of the interferometer using the holey mirror. The pump and probe beams are then focused into the interaction region with a toroidal mirror. A schematic of the experimental setup is shown in Fig. 6.1.

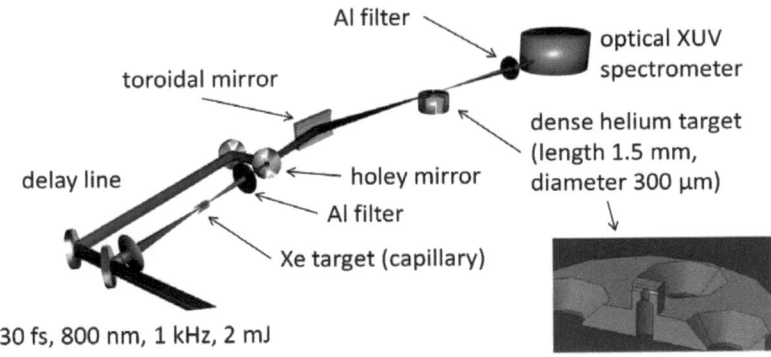

Fig. 6.1 Schematic of the experimental setup for the attosecond transient absorption measurement.

The harmonic radiation transmitted through the interaction gas target is refocused and detected in the XUV spectrometer. The IR from the probe beam is removed before the spectrometer by a second aluminum filter. This is necessary because the CCD is sensitive in the IR spectral region.

For the experiment we have chosen helium as target medium. Its ionization potential is lower than the photon energy of harmonic order 17. Harmonic photons with order ≥ 17 are thus absorbed in a single photon process. With the target used for the TOF measurements, the gas density is too low to observe absorption of the harmonic radiation in the XUV spectrometer. The gas load in a TOF measurement needs to be low, such that the mean free path is large enough for the electron propagation. Furthermore the TOF detector is equipped with a MCP, which requires pressures in the range of 10^{-6} mbar. For the absorption measurement, we are not limited by the electron detector anymore which permits an increase of the density in the interaction region. This is achieved by replacing the TOF target (Fig. 4.18) with the pulsed valve (Sec. 4.3.2.1). We installed the "T"-shaped target tip (Fig. 4.12) to achieve highest gas density. The gas pressure of this target was adjusted such that about half of the photons of harmonic 17 were absorbed. From that we estimate the density in the target to be $5 \cdot 10^{17}$ particles/cm^3 [71].

Radiation of harmonic order 15 and less might be absorbed in the target medium if there is overlap of the harmonic spectrum with atomic levels of helium. However, ionization of helium by these low energetic harmonics occurs only in a multiphoton process. As the intensity of and two-photon cross-sections for the harmonics are low [23], this is only possible in the presence of an additional and suitably strong IR field.

Before we discuss the experimental results, some properties of the harmonic emission from xenon have to be mentioned: The focus was positioned at the exit of the capillary target. The laser thus propagates through xenon before the high harmonics are generated. The pressure in the target was 24 mbar. The intensity was large enough to ionize xenon and induce a significant blue-shift of the driving laser field before the generation (see also Fig. 3c in [84]). The fundamental photon energy in the generation is estimated to be 1.58 eV. In the probe arm the photon energy is not changed and is 1.55 eV.

6.2 Absorption of high harmonics for large delays

In this section we will first investigate the situation where there is no temporal overlap between the XUV and IR fields. We will measure the transmission behavior for following three situations:

i. Without helium in the interaction region. This allows measuring the yield that was generated in xenon.

ii. With helium and the IR pulse 60 fs after the XUV. At this delay the absorption behavior cannot be influenced by the IR field.

iii. With helium and the IR pulse 60 fs before the XUV. At this delay the absorption behavior could be changed by the strong IR pulse if there is a non-instantaneous coupling.

In Fig. 6.2 the spectrum of harmonic order 17 is shown for the three cases. One can observe a clear absorption when the helium is present in the interaction region. The behavior is identical for cases ii. and iii. The ionization potential of helium is 24.587 eV. The photons contained in harmonic 17 have an energy that is well above the ionization potential, the absorption happens into the continuum state in a single photon process. The absorption behavior is not changed, even if an IR field arrives 60 fs before the harmonic, as can be seen in the experimental data.

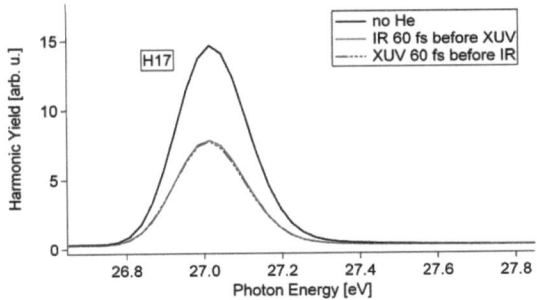

Fig. 6.2 Close up of harmonic 17. Signal without helium in the interaction (black), signal with helium where the IR pulse is 60 fs before the XUV (red) and 60 fs after the XUV (blue). The IR intensity is $1.3 \cdot 10^{13}$ W cm^{-2}.

The situation is completely different for harmonic 15. It has a photon energy ranging from 23.6 eV to 24.1 eV which is below, but close to the ionization potential of helium. The spectrum is shown for the three cases in Fig. 6.3. We can identify two strong absorption peaks corresponding to the 1s4p and 1s5p levels of helium (see Appendix A.1) when looking at the relative absorption (signal without helium divided by the signal with helium) which is shown in Fig. 6.4 for case ii.

This absorption structure is even more pronounced when the IR field arrives 60 fs before the XUV. The IR pulse has an intensity of $1.3 \cdot 10^{13}$ W cm^{-2}. It seems to increase the absorption cross section, especially of the 1s4p level. The induced change is persistent for at least 60 fs. We will further investigate this behavior in the situation where the two fields overlap in time in section 6.6.

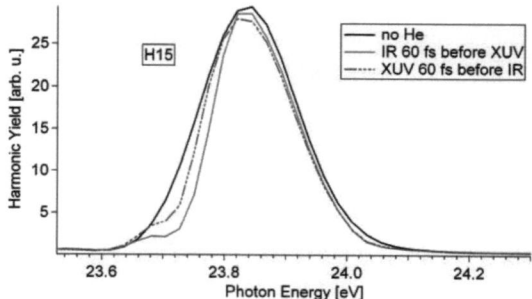

Fig. 6.3 Close up of harmonic 15, same conditions as in Fig. 6.2.

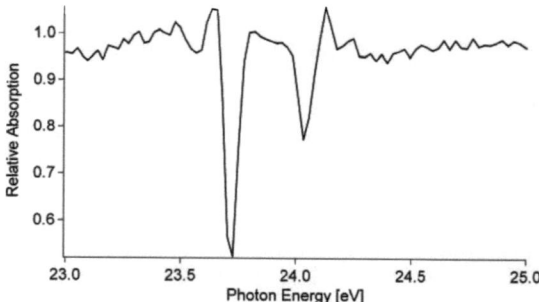

Fig. 6.4 Relative absorption of harmonic 15, i.e. signal without helium divided by the signal with helium from Fig. 6.3 for the case where the XUV arrives 60 fs before the IR.

6.3 Typical delay scan

When moving the XUV/IR delay across the overlap region, new structures appear in the spectrum. In Fig. 6.5 a typical delay scan is shown for an IR intensity of $1.3 \cdot 10^{13}$ W cm^{-2}. For better visibility, the harmonic signal has been plotted on a logarithmic scale. This is the only graph in logarithmic scale, all other results are shown in linear scale.

Harmonic orders 13 and 15 show a significant absorption which is controlled by the time delay and the intensity of the strong IR field of the laser. This structure is on a femtosecond timescale and will be discussed in Sec. 6.4. In addition, a structure on the attosecond timescale was measured. It will be described in Sec. 6.5.

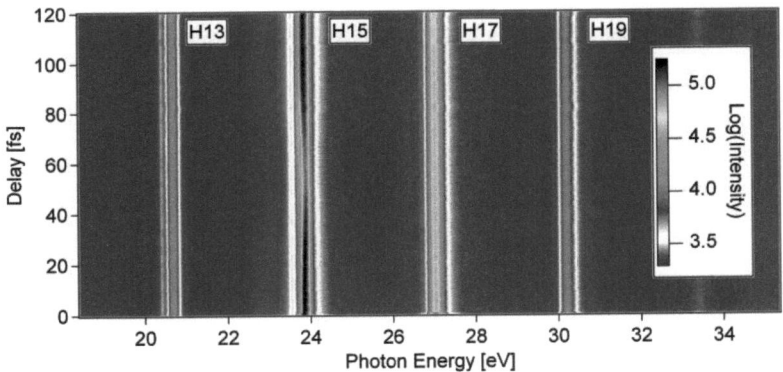

Fig. 6.5 Typical delay scan in logarithmic scale for better visibility. The infrared intensity was $1.3 \cdot 10^{13}$ W cm^{-2}.

To acquire a complete delay scan of 120 fs delay range with 1121 delay-steps, i.e. single steps of 107 as, approximately 6 minutes of measurement time are required. This is far less than for the usual electron or ion detection as used in the RABBITT measurement, where a complete scan usually takes several hours.

6.4 Femtosecond structure

On the femtosecond timescale, a clear envelope structure becomes visible when performing a spectral integration of harmonic order 13 and 15 as shown in Fig. 6.6.

The FWHM of the cross correlation is 43 fs for harmonic 15 and 23 fs for harmonic 13. This indicates that the nonlinearity for the absorption of harmonic 13 is much higher than for the case of harmonic 15, which is expected because harmonic 15 is close to the threshold. A single photon from the IR field is sufficient to ionize helium in combination with harmonic 15. For harmonic 13 it requires 3 IR photons for ionization.

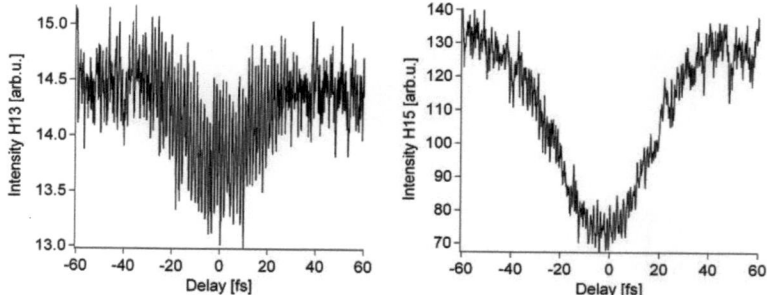

Fig. 6.6 Spectrally integrated signal for harmonic 13 (left) and harmonic 15 (right) over the full spectrum of the harmonic. A clear structure on a fs timescale becomes visible. On top of it a faster modulation on an attosecond timescale is visible (see Sec. 6.5). The infrared intensity was $1.3 \cdot 10^{13}$ W cm^{-2}.

In contrast to the strong increase in absorption of harmonic 13 and 15, the absorption of harmonic 17 which has a photon energy higher than the ionization potential of helium stays unchanged. The spectrally integrated signal is shown in Fig. 6.7.

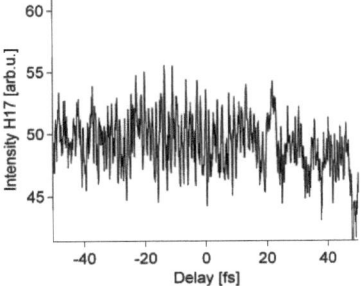

Fig. 6.7 Spectrally integrated signal of harmonic 17 which has a photon energy higher than the ionization potential of helium. In this case, the absorption is not significantly changed on a femtosecond timescale. The infrared intensity was $1.3 \cdot 10^{13}$ W cm^{-2}.

6.5 Attosecond structure

In addition to the femtosecond structure, a sub-cycle modulation of the transmission (with half-cycle periodicity of the fundamental laser frequency) appears. This modulation is visible on all harmonics, even for those with photon energy higher than the ionization potential of helium. The absorption cross section of harmonic 17 is clearly modulated. These fast oscillations are already visible in Fig. 6.6 and Fig. 6.7. In this section we will further investigate the properties of these oscillations.

6.5.1 Phase differences in the modulation of the transmission of individual harmonics

Fig. 6.8 shows the separately spectrally integrated yield of harmonic 13, 15 and 17 for a laser intensity of $1.3 \cdot 10^{13}$ W cm^{-2} in the probe beam. We find that harmonic 13 and 17 are modulating nearly in phase. More precisely, we see a phase difference between 0 rad and -0.23 rad in repetitive scans that we measured. In comparison to this there is a large phase difference between harmonic 13 and 15 ranging from 2.15 to 2.55 rad in our experiment.

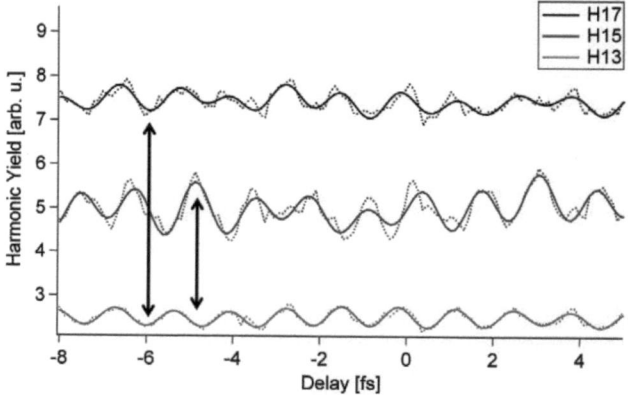

Fig. 6.8 Oscillations of the individual harmonic yield (orders 13, 15 and 17) around zero delay. The signal has been shifted vertically such that the harmonics appear underneath each other for better comparison of the relative phase differences. Data points are shown as dotted line, the solid line is a Fourier filtered signal, where oscillations faster than twice the fundamental laser frequency have been removed. The measurement has been taken with an aluminum filter thickness of 500 nm and an IR probe intensity of $1.3 \cdot 10^{13}$ W cm^{-2}.

In a next step we tried to investigate whether these relative phase differences are related to the relative phase of the harmonics in the APT. This phase corresponds to the "atto-chirp" which can be determined in a RABBITT measurement. It can be changed by changing the thickness of the aluminum filter used after the generation [78]. When changing the thickness of the aluminum filter after the generation from 500 nm to 100 nm we calculate a relative phase shift between harmonic 13 and 15 of approximately 1.3 rad, corresponding to 280 as. For harmonic 13 and 17 we expect a change of 1.74 rad or 375 as. This calculation is based on the refractive index data from [79]. In our measurement, however, we did not observe any relative phase shift of the modulated transmission resulting from a change in filter thickness. This leads us to conclude that the recorded transient absorption is not directly depending on the temporal structure of the pulses in the APT. This result is surprising because the half-cycle modulation can only be explained by having an APT.

We also performed measurements at a higher IR probe intensity of $1.9 \cdot 10^{13}$ W cm^{-2}. In contrast to varying the "atto-chirp" of the xenon APT, we find that this intensity change has a strong effect on the relative phase between the absorption modulations of individual harmonics. For this higher intensity, we find a phase difference between harmonic 13 and 15 in the range of 3.5 to 4 rad. These harmonics are still oscillating nearly out of phase. However, at the lower probe intensity the peaks in harmonic 13 were before the peaks of harmonic 15, whereas here harmonic 15 peaks before harmonic 13. Harmonic 13 and 17 are now also nearly out of phase (phase difference ranging from -1.7 to -2.8), whereas for the lower probe intensity harmonic 13 and 17 were in phase.

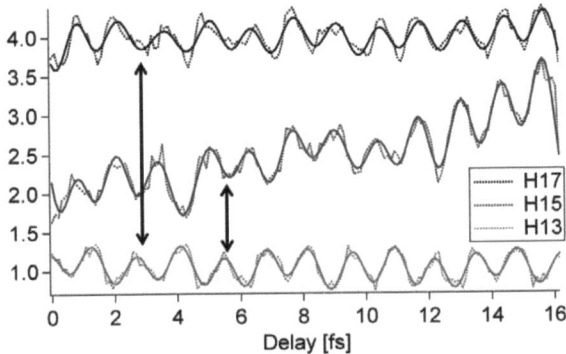

Fig. 6.9 Oscillations of individual harmonics (orders 13, 15 and 17) around zero delay. The signal has been shifted such that the harmonics appear underneath each other for better comparison of the relative phase differences. Data points are shown as dotted line, the solid line is a Fourier filtered signal, where oscillations faster than twice the fundamental laser frequency have been removed. The measurement has been taken with an aluminum filter thickness of 500 nm and an IR probe intensity of $1.9 \cdot 10^{13}$ W cm^{-2}.

The transmitted photon yield at harmonics 13 through 17 for the higher probe intensity is shown in Fig. 6.9. For this intensity value, we repeated the measurements for the same two Aluminum filter thicknesses listed above to confirm that the observed phase differences are indeed independent of the "atto-chirp".

6.5.2 Total transmitted photon yield

In a recent experiment performed by P. Johnsson and co-workers [31], it was shown that the ion yield from helium modulates with twice the fundamental laser frequency when measuring a delay scan in experimental conditions similar to ours. This experiment allowed to measure the total ionization probability in the combined field since it is proportional to the recorded ion yield signal. In their paper they also performed numerical calculations based on the integration of the TDSE with the finding that not only the ionization probability oscillates with XUV/IR delay but also the absorption probability will show this oscillation. The latter is a parameter we have experimentally access to. Note that the ionization is not directly linked to the absorption, because the absorption can also lead to an excited state in helium without ionizing it.

To extract the total transmitted photon yield from our data, we calculated the total harmonic signal by spectral integration. The result is shown in Fig. 6.10. A clear modulation of the harmonic yield with twice the fundamental laser frequency is visible which is in agreement with [31]. We did this calculation at both laser intensities used in our experiment. The modulation of the transmitted yield is present in both cases.

Due to the experimental setup, the measured spectra are not directly linked to the total absorption probability by a spectral integration. The modulation of the total yield could be an effect of a poorly calibrated spectral sensitivity of our detection system, which might simply pronounce one harmonic. We can rule this out by analyzing the oscillations on individual harmonics separately as we did in the previous section. The relative phase behavior of these modulations cannot cancel in the total yield because we find components that are oscillating in phase, but no component that is oscillating exactly out of phase. This means that the individual signals cannot cancel each other in the summation, even not when changing their relative amplitudes.

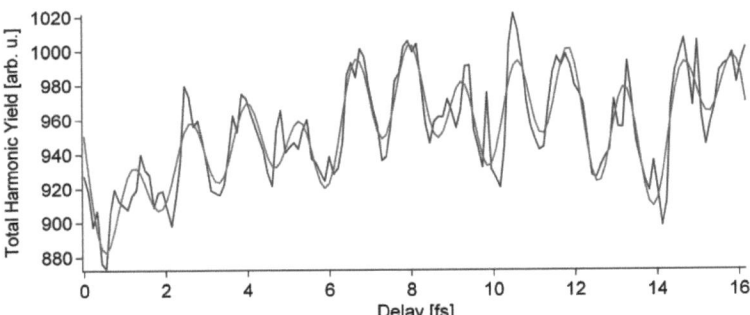

Fig. 6.10 A spectral integration of all the harmonics. The total yield shows an oscillation with half-cycle periodicity. The blue line corresponds to the experimental data, the red line is a smoothed version of the data (Fourier filtered).

To further compare our results to the measurements taken in [31], we have performed scans with argon as target medium where the photon energies of all harmonics are well above the ionization potential. In this case only a static absorption with no modulation of the transmitted harmonics has been measured. The same was found in [31] for the ion yield.

6.6 Asymmetric structure in the absorption spectra of harmonic 15

We recorded transmission spectra in the absence of helium in the interaction region for each scan, as shown for large delays in Fig. 6.2 and Fig. 6.3. It allowed calculating a relative absorption by dividing the signal without helium by the signal with helium (see Fig. 6.4). We identified the 1s4p and 1s5p absorption lines of helium in harmonic 15. We have already seen that the absorption knee of the 1s4p line becomes more pronounced when the IR pulse arrives 60 fs before the XUV. Here, we investigate the delay dependent behavior of this structure in the overlap region of the XUV and IR radiation.

In Fig. 6.11 the relative absorption is plotted for harmonic 15 for a complete delay scan. The IR intensity was $1.3 \cdot 10^{13}$ W cm^{-2}. The signal shown in blue corresponds to absorption, the one in red to enhancement. In the overlap region, a clear broadening of the 1s4p absorption line is measured. The 1s5p absorption is much weaker and less affected. As already discussed in Sec. 6.2, the absorption is changed even for a delay of -60 fs, where the IR arrives before the XUV and the pulses have very weak overlap only.

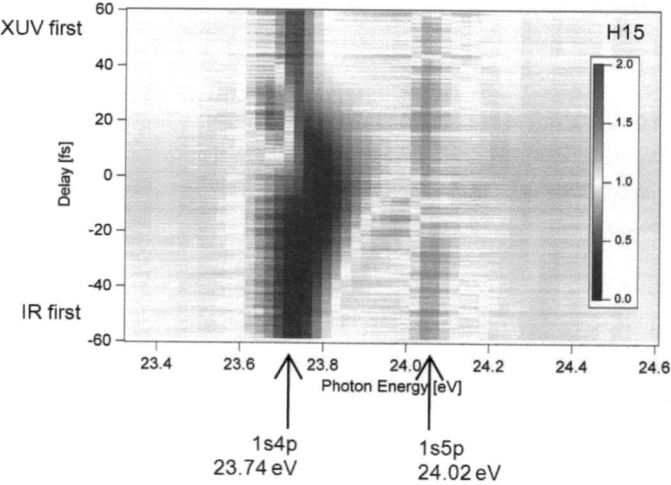

Fig. 6.11 Relative absorption of harmonic 15, i.e. signal without helium divided by the signal with helium for the full delay scan. The absorption is strongest for zero time delay. At a delay of 20 fs (XUV before IR) a localized enhancement of the signal is measured. The enhancement factor reaches the value of two. The IR intensity was $1.3 \cdot 10^{13}$ W cm^{-2}.

For a delay of 20 fs, where the XUV arrives before the IR, a localized enhancement of the signal is detected. This enhancement reaches a factor of up to two. It oscillates with half-cycle periodicity like the absorption. It becomes weaker as the IR probe intensity is decreased. We performed measurements with an IR intensity of $1.1 \cdot 10^{13}$ W cm^{-2} where the structure is still visible. At an intensity of $0.7 \cdot 10^{13}$ W cm^{-2} it is not detected anymore.

In contrast to harmonic 15, the relative absorption of harmonic order 17 is constant (besides the modulation with half-cycle periodicity) over the complete delay range as shown in Fig. 6.12. In harmonic 13 we didn't find any asymmetric behavior of the absorption, either.

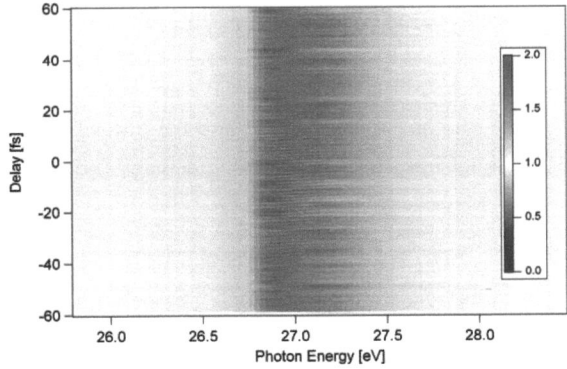

Fig. 6.12 Relative absorption of harmonic 17, i.e. signal without helium divided by the signal with helium for the full delay scan. The average absorption is constant over the complete delay range. The IR intensity was $1.3 \cdot 10^{13}$ W cm^{-2}.

6.7 Theoretical description of the experiment

The theoretical treatment of our experimental scenario is a challenging task. The atomic levels of helium have to be taken into account, thus strong field approximation cannot be used for this purpose. Implementing absorption of photons in TDSE is very uncommon. Theoreticians typically calculate the generated electrons or ions and their energy distribution. To resolve which photon created these electrons requires a new approach. The theoretical treatment of the experiment just started and is performed by the group of Prof. J. M. Rost in Dresden, Germany.

6.8 Conclusion and outlook

In the attosecond transient absorption experiment the transmission of an APT generated in xenon through a dense helium target was measured. The absorption was influenced by a strong IR field. We found structures on the femtosecond timescale: the envelope structure of the IR/XUV cross correlation was measured in harmonic 13 and 15. A localized enhancement of the signal in harmonic 15 was found. In addition the experiment shows a structure on the attosecond timescale: The transmitted photon yield oscillates with half-cycle periodicity of the driving field. We investigated this feature by changing the IR intensity of the driving field and the "atto-chirp" of the APT. Our results of the total transmission are in agreement with the previously published work of J. Mauritsson and co-workers.

Our experiment was measured in an all-optical scheme, thus not involving the typical ion or electron detection in attosecond experiments. This results in a very fast data acquisition and a good signal-to-noise ratio.

Currently the data is theoretically investigated by the group of Prof. J. M. Rost. His group is working on a simulation of the absorption by numerically solving TDSE.

When performing the experiment with an isolated attosecond pulse, an even richer structure is expected since the continuous spectrum of these pulses will have overlap with more levels in the helium atom.

We hope that this measurement will allow extracting detailed information on the target medium and the absorption dynamics close to the ionization threshold. This would enable the technique to be extended to other, more complex target media.

Chapter 7

Summary And Outlook

In this thesis, high harmonics generated by a low frequency laser field in noble gases have been investigated in both, spectral and temporal domain.

Experimental conditions have been found allowing for the observation of a spectral interference structure between the two shortest electron quantum paths of the harmonic generation process. This interference revealed itself as a modulation of the harmonic yield when increasing the intensity of the generating laser pulse. It was possible to measure QPI in different generation media: neon, argon and xenon. This allowed investigating QPI with laser intensities below, around and above the intensity of barrier suppression. Ionization effects have become visible in the QPI structures. The experimental results are in good agreement with theoretical simulations performed by T. Auguste and co-workers. It was possible to also spatially resolve QPI for a generation in argon.

QPI and more precisely, the electron excursion times can be controlled by the strong laser field with attosecond resolution: at a laser intensity of $1.5 \cdot 10^{14}$ W cm^{-2}, a variation of 10% in intensity shifts the QPI from constructive to destructive interference for harmonic 15 in argon. This corresponds to a relative change in the timing of the electron quantum paths of 35 as only.

Through the interference structures, the experimental method gives direct access to the electronic properties of the HHG process. The scheme provides the possibility of investigating QPI in more complex systems such as molecules. The interferometric sensitivity of QPI should reveal differences between the molecular and the atomic continua.

The experiment on QPI was purely in spectral domain. It didn't use the attosecond temporal structure that can be obtained from HHG. When generating harmonics with a multi-cycle driving field, an APT can be generated. It is inherently synchronized to the driving electric laser field. This allows for XUV/IR pump-probe measurements with attosecond resolution. For these experiments attosecond mechanical stability in optical delay lines is required, which corresponds to a change in relative path length of a few nanometers. To enable such measurements, a new harmonic beamline and a field-free electron time-of-flight spectrometer for attosecond experiments have been designed and built.

The first experiment with attosecond resolution performed in this beamline was the complete characterization of an APT generated in argon. The average pulse duration has been measured to be 450 as using the RABBITT technique. The flux of the harmonics was 1.4 nJ per laser shot. The beam profile of the harmonics was close to Gaussian. An APT in the low harmonic spectral range (harmonic 13 to 19) has been generated in xenon. The RABBITT measurement indicates an average pulse duration of 400 as.

This APT has been applied in an attosecond transient absorption experiment. The harmonics have been absorbed in a dense helium target and the transmitted photon yield has been detected in an XUV spectrometer. The absorption was influenced by a strong IR field in an XUV/IR pump-probe measurement. The transmission not only shows a structure on the femtosecond timescale, but also on the attosecond timescale. This measurement is a first step towards all-optical attosecond experiments. The photon detection results in a good signal to noise ratio. The data acquisition is faster compared to the usual experiments based on the detection of photoelectrons or –ions.

The absorption measurements allowed identifying the 1s4p and 1s5p levels in helium and their dynamic behavior with XUV/IR pump-probe delay. In addition, a localized enhancement of the signal was found for some delays.

The experimental findings are currently investigated by means of a theoretical simulation by the group of Prof. J.-M. Rost, Dresden, Germany. A successful numerical reconstruction might allow applying the absorption scheme to more complex systems such as molecules and extracting information on the involved electron dynamics. It could be interesting to perform the measurements with isolated attosecond pulses. Compared to the odd harmonics of an APT the continuous spectrum of these pulses has more overlap with the atomic levels. This enables a measurement of the dynamic absorption structure in a continuous spectral range.

References

[1] T. H. Maiman, "Stimulated Optical Radiation in Ruby," *Nature,* vol. 187, pp. 493-494, 1960.

[2] A. H. Zewail, "Laser Femtochemistry," *Science,* vol. 242, pp. 1645-1653, Dec 1988.

[3] A. H. Zewail, "Femtochemistry: Atomic-scale dynamics of the chemical bond," *Journal of Physical Chemistry A,* vol. 104, pp. 5660-5694, Jun 2000.

[4] T. Udem, R. Holzwarth, and T. W. Hansch, "Optical frequency metrology," *Nature,* vol. 416, pp. 233-237, Mar 2002.

[5] A. Marian, M. C. Stowe, J. R. Lawall, D. Felinto, and J. Ye, "United time-frequency spectroscopy for dynamics and global structure," *Science,* vol. 306, pp. 2063-2068, Dec 2004.

[6] A. McPherson, G. Gibson, H. Jara, U. Johann, T. S. Luk, I. A. McIntyre, K. Boyer, and C. K. Rhodes, "Studies of multiphoton production of vacuum ultraviolet-radiation in the rare-gases," *Journal of the Optical Society of America B-Optical Physics,* vol. 4, pp. 595-601, Apr 1987.

[7] M. Ferray, A. L'Huillier, X. F. Li, L. A. Lompre, G. Mainfray, and C. Manus, "Multiple-harmonic conversion of 1064-nm radiation in rare-gases," *Journal of Physics B-Atomic Molecular and Optical Physics,* vol. 21, pp. L31-L35, Feb 1988.

[8] P. Salieres, A. L'Huillier, P. Antoine, and M. Lewenstein, "Study of the spatial and temporal coherence of high-order harmonics," *Adv. At. Mol. Opt. Phys.,* vol. 41, pp. 83-142, 1999.

[9] P. Antoine, A. L'Huillier, and M. Lewenstein, "Attosecond pulse trains using high-order harmonics," *Physical Review Letters,* vol. 77, pp. 1234-1237, Aug 1996.

[10] E. Goulielmakis, M. Schultze, M. Hofstetter, V. S. Yakovlev, J. Gagnon, M. Uiberacker, A. L. Aquila, E. M. Gullikson, D. T. Attwood, R. Kienberger, F. Krausz, and U. Kleineberg, "Single-Cycle Nonlinear Optics," *Science,* vol. 320, pp. 1614-1617, 2008.

[11] K. C. Kulander, K. J. Schafer, and J. L. Krause, "Dynamics of short-pulse excitation, ionization and harmonic conversion," in *NATO Advanced Research Workshop on SILAP (Super-Intense Laser-Atom Physics),* Han-Sur-Lesse, Belgium, 1993, pp. 95-110.

[12] P. B. Corkum, "Plasma Perspective on Strong-Field Multiphoton Ionization," *Phys. Rev. Lett.,* vol. 71, pp. 1994-1997, 1993.

[13] E. Seres, J. Seres, and C. Spielmann, "X-ray absorption spectroscopy in the keV range with laser generated high harmonic radiation," *Applied Physics Letters,* vol. 89, p. 181919, Oct 2006.

[14] M. Lewenstein, P. Balcou, M. Y. Ivanov, A. L'Huillier, and P. B. Corkum, "Theory of high-harmonic generation by low-frequency laser fields," *Phys. Rev. A,* vol. 49, pp. 2117-2132, 1994.

[15] P. Salieres, B. Carre, L. Le Deroff, F. Grasbon, G. G. Paulus, H. Walther, R. Kopold, W. Becker, D. B. Milosevic, A. Sanpera, and M. Lewenstein, "Feynman's path-integral approach for intense-laser-atom interactions," *Science,* vol. 292, pp. 902-905, May 2001.

[16] M. Lewenstein, P. Salieres, and A. L'Huillier, "Phase of the atomic polarization in high-order harmonic generation," *Phys. Rev. A,* vol. 52, pp. 4747-4754, 1995.

[17] J. Mauritsson, P. Johnsson, R. Lopez-Martens, K. Varju, W. Kornelis, J. Biegert, U. Keller, M. B. Gaarde, K. J. Schafer, and A. L'Huillier, "Measurement and control of the frequency chirp rate of high-order harmonic pulses," *Physical Review A,* vol. 70, p. 021801, Aug 2004.

[18] G. Sansone, E. Benedetti, J. P. Caumes, S. Stagira, C. Vozzi, S. De Silvestri, and M. Nisoli, "Control of long electron quantum paths in high-order harmonic generation by phase-stabilized light pulses," *Physical Review A,* vol. 73, p. 053408, May 2006.

[19] P. Balcou, P. Salieres, A. L'Huillier, and M. Lewenstein, "Generalized phase-matching conditions for high harmonics: The role of field-gradient forces," *Physical Review A,* vol. 55, pp. 3204-3210, Apr 1997.

[20] B. Walker, B. Sheehy, L. F. DiMauro, P. Agostini, K. J. Schafer, and K. C. Kulander, "Precision Measurement of Strong Field Double Ionization of Helium," *Physical Review Letters,* vol. 73, p. 1227, 1994.

[21] D. Strickland and G. Mourou, "Compression of amplified chirped optical pulses," *Optics Communications,* vol. 56, pp. 219-221, 1985.

[22] P. Tzallas, D. Charalambidis, N. A. Papadogiannis, K. Witte, and G. D. Tsakiris, "Direct observation of attosecond light bunching," *Nature,* vol. 426, pp. 267-271, Nov 2003.

[23] L. A. A. Nikolopoulos and P. Lambropoulos, "Multichannel theory of two-photon single and double ionization of helium," *Journal of Physics B-Atomic Molecular and Optical Physics,* vol. 34, pp. 545-564, 2001.

[24] C. Lynga, M. B. Gaarde, C. Delfin, M. Bellini, T. W. Hansch, A. L'Huillier, and C. G. Wahlstrom, "Temporal coherence of high-order harmonics," *Physical Review A,* vol. 60, pp. 4823-4830, Dec 1999.

[25] P. M. Paul, E. S. Toma, P. Breger, G. Mullot, F. Augé, P. Balcou, H. G. Muller, and P. Agostini, "Observation of a Train of Attosecond Pulses from High Harmonic Generation," *Science,* vol. 292, pp. 1689-1692, 2001.

[26] M. Drescher, M. Hentschel, R. Kienberger, G. Tempea, C. Spielmann, G. A. Reider, P. B. Corkum, and F. Krausz, "X-ray pulses approaching the attosecond frontier," *Science,* vol. 291, pp. 1923-1927, Mar 2001.

[27] E. Goulielmakis, M. Uiberacker, R. Kienberger, A. Baltuska, V. Yakovlev, A. Scrinzi, T. Westerwalbesloh, U. Kleineberg, U. Heinzmann, M. Drescher, and F. Krausz, "Direct measurement of light waves," *Science,* vol. 305, pp. 1267-1269, Aug 27 2004.

[28] M. Drescher, M. Hentschel, R. Kienberger, M. Uiberacker, V. Yakovlev, A. Scrinzi, T. Westerwalbesloh, U. Kleineberg, U. Heinzmann, and F. Krausz, "Time-resolved atomic inner-shell spectroscopy," *Nature,* vol. 419, pp. 803-807, Oct 2002.

[29] J. Itatani, J. Levesque, D. Zeidler, H. Niikura, H. Pepin, J. C. Kieffer, P. B. Corkum, and D. M. Villeneuve, "Tomographic imaging of molecular orbitals," *Nature,* vol. 432, pp. 867-871, Dec 2004.

[30] R. Torres, N. Kajumba, J. G. Underwood, J. S. Robinson, S. Baker, J. W. G. Tisch, R. de Nalda, W. A. Bryan, R. Velotta, C. Altucci, I. C. E. Turcu, and J. P. Marangos, "Probing orbital structure of polyatomic molecules by high-order harmonic generation," *Physical Review Letters,* vol. 98, p. 203007, Jun 2007.

[31] P. Johnsson, J. Mauritsson, T. Remetter, A. L'Huillier, and K. J. Schafer, "Attosecond Control of Ionization by Wave-Packet Interference," *Physical Review Letters,* vol. 99, p. 233001, 2007.

[32] L. V. Keldysh, "Ionization in the field of a strong electromagnetic wave," *Sov. Phys. JETP,* vol. 20, p. 1307, 1965.

[33] S. Augst, D. Strickland, D. D. Meyerhofer, S. L. Chin, and J. H. Eberly, "Tunneling ionization of noble gases in a high-intensity laser field," *Physical Review Letters,* vol. 63, p. 2212, 1989.

[34] M. W. Walser, C. H. Keitel, A. Scrinzi, and T. Brabec, "High harmonic generation beyond the electric dipole approximation," *Physical Review Letters,* vol. 85, pp. 5082-5085, Dec 2000.

[35] J. L. Krause, K. J. Schafer, and K. C. Kulander, "High-Order Harmonic generation from Atoms and Ions in the High Intensity Regime," *Phys. Rev. Lett.*, vol. 68, pp. 3535-3538, 1992.

[36] I. P. Christov, J. Zhou, J. Peatross, A. Rundquist, M. M. Murnane, and H. C. Kapteyn, "Nonadiabatic effects in high-harmonic generation with ultrashort pulses," *Physical Review Letters*, vol. 77, pp. 1743-1746, Aug 1996.

[37] P. Balcou, A. S. Dederichs, M. B. Gaarde, and A. L'Huillier, "Quantum-path analysis and phase matching of high-order harmonic generation and high-order frequency mixing processes in strong laser fields," *Journal of Physics B-Atomic Molecular and Optical Physics*, vol. 32, pp. 2973-2989, Jun 28 1999.

[38] N. H. Burnett, C. Kan, and P. B. Corkum, "Ellipticity and polarization effects in harmonic-generation in ionizing neon," *Physical Review A*, vol. 51, pp. R3418-R3421, May 1995.

[39] P. Antoine, A. L'Huillier, M. Lewenstein, P. Salieres, and B. Carre, "Theory of high-order harmonic generation by an elliptically polarized laser field," *Physical Review A (Atomic, Molecular, and Optical Physics)*, vol. 53, pp. 1725-1745, March 1996.

[40] K. Burnett, V. C. Reed, J. Cooper, and P. L. Knight, "Calculation of the background emitted during high-harmonic generation," *Physical Review A*, vol. 45, pp. 3347-3349, Mar 1992.

[41] R. P. Feynman, "Space-Time Approach to Non-Relativistic Quantum Mechanics," *Reviews of Modern Physics*, vol. 20, pp. 367-387, 1948.

[42] L. G. Gouy, "Sur une propriete nouvelle des ondes lumineuses," *Comptes Rendus de l'Académie des Sciences*, vol. 110, p. 1251, 1890.

[43] A. L'Huillier, P. Balcou, S. Candel, K. J. Schafer, and K. C. Kulander, "Calculations of High-Order Harmonic-Generation Processes in Xenon at 1064 Nm," *Physical Review A*, vol. 46, pp. 2778-2790, Sep 1 1992.

[44] S. C. Rae and K. Burnett, "Detailed Simulations of Plasma-Induced Spectral Blueshifting," *Physical Review A*, vol. 46, pp. 1084-1090, Jul 15 1992.

[45] T. Auguste, P. Monot, L. A. Lompre, G. Mainfray, and C. Manus, "Defocusing Effects of a Picosecond Terawatt Laser-Pulse in an Underdense Plasma," *Optics Communications*, vol. 89, pp. 145-148, May 1 1992.

[46] P. Salieres, A. L'Huillier, and M. Lewenstein, "Coherence Control of High-Order Harmonics," *Physical Review Letters*, vol. 74, pp. 3776-3779, May 1995.

[47] G. Farkas and C. Toth, "Proposal for attosecond light-pulse generation using laser-induced multiple-harmonic conversion processes in rare-gases," *Physics Letters A*, vol. 168, pp. 447-450, Sep 1992.

[48] P. F. Moulton, "Spectroscopic and laser characteristics of Ti:Al$_2$O$_3$," *J. Opt. Soc. Am. B*, vol. 3, pp. 125-132, 1986.

[49] D. E. Spence, P. N. Kean, and W. Sibbett, "60-fsec pulse generaion from a self-mode-locked Ti-Sapphire laser," *Optics Letters*, vol. 16, pp. 42-44, Jan 1991.

[50] D. H. Sutter, L. Gallmann, N. Matuschek, F. Morier-Genoud, V. Scheuer, G. Angelow, T. Tschudi, G. Steinmeyer, and U. Keller, "Sub-6-fs pulses from a SESAM-assisted Kerr-lens modelocked Ti : sapphire laser: at the frontiers of ultrashort pulse generation," *Applied Physics B-Lasers and Optics*, vol. 70, pp. S5-S12, Jun 2000.

[51] U. Morgner, F. X. Kartner, S. H. Cho, E. Chen, H. A. Haus, J. G. Fujimoto, E. P. Ippen, V. Scheuer, G. Angelow, and T. Tschudi, "Sub-

two-cycle pulses from a Kerr-lens mode-locked Ti : sapphire laser," *Optics Letters*, vol. 24, pp. 920-920, Jul 1 1999.

[52] F. Verluise, V. Laude, Z. Cheng, C. Spielmann, and P. Tournois, "Amplitude and phase control of ultrashort pulses by use of an acousto-optic programmable dispersive filter: pulse compression and shaping," *Opt. Lett.*, vol. 25, pp. 575-577, 2000.

[53] M. Nisoli, S. DeSilvestri, and O. Svelto, "Generation of high energy 10 fs pulses by a new pulse compression technique," *Applied Physics Letters*, vol. 68, pp. 2793-2795, May 13 1996.

[54] C. P. Hauri, W. Kornelis, F. W. Helbing, A. Heinrich, A. Courairon, A. Mysyrowicz, J. Biegert, and U. Keller, "Generation of intense, carrier-envelope phase-locked few-cycle laser pulses through filamentation," *Appl. Phys. B*, vol. 79, pp. 673-677, 2004.

[55] A. Guandalini, P. Eckle, M. P. Anscombe, P. Schlup, J. Biegert, and U. Keller, "5.1 fs pulses generated by filamentation and carrier envelope phase stability analysis," *J. Phys. B: At. Mol. Opt. Phys.*, vol. 39, pp. S257-S264, 2006.

[56] K. J. Schafer, B. Yang, L. F. Dimauro, and K. C. Kulander, "Above Threshold Ionization Beyond The High Harmonic Cutoff," *Physical Review Letters*, vol. 70, pp. 1599-1602, Mar 1993.

[57] J. J. Macklin, J. D. Kmetec, and C. L. Gordon, "High-order harmonic-generation using intense femtosecond pulses," *Physical Review Letters*, vol. 70, pp. 766-769, Feb 1993.

[58] C. G. Wahlstrom, J. Larsson, A. Persson, T. Starczewski, S. Svanberg, P. Salieres, P. Balcou, and A. L'Huillier, "High-Order Harmonic Generation in Rare-Gases with an Intense Short-Pulse-Laser," *Physical Review A*, vol. 48, pp. 4709-4720, Dec 1993.

[59] E. Benedetti, J. P. Caumes, G. Sansone, S. Stagira, C. Vozzi, and M. Nisoli, "Frequency chirp of long electron quantum paths in high-order harmonic generation," *Optics Express*, vol. 14, pp. 2242-2249, Mar 2006.

[60] C. Corsi, A. Pirri, E. Sali, A. Tortora, and M. Bellini, "Direct interferometric measurement of the atomic dipole phase in high-order harmonic generation," *Physical Review Letters*, vol. 97, p. 023901, Jul 2006.

[61] H. Merdji, M. Kovacev, W. Boutu, P. Salieres, F. Vernay, and B. Carre, "Macroscopic control of high-order harmonics quantum-path components for the generation of attosecond pulses," *Physical Review A*, vol. 74, p. 043804, Oct 2006.

[62] Z. Chang, A. Rundquist, H. Wang, I. Christov, H. C. Kapteyn, and M. M. Murnane, "Temporal phase control of soft-x-ray harmonic emission," *Physical Review A*, vol. 58, pp. R30-R33, Jul 1998.

[63] M. B. Gaarde, F. Salin, E. Constant, P. Balcou, K. J. Schafer, K. C. Kulander, and A. L'Huillier, "Spatiotemporal separation of high harmonic radiation into two quantum path components," *Physical Review A*, vol. 59, pp. 1367-1373, Feb 1999.

[64] T. Auguste, P. Salieres, A. S. Wyatt, A. Monmayrant, I. A. Walmsley, E. Cormier, A. Zair, M. Holler, A. Guandalini, F. Schapper, J. Biegert, L. Gallmann, and U. Keller, "Theoretical and experimental analysis of quantum path interferences in high-order harmonic generation," *Physical Review A*, vol. 80, p. 033817, 2009.

[65] H. R. Telle, G. Steinmeyer, A. E. Dunlop, J. Stenger, D. H. Sutter, and U. Keller, "Carrier-envelope offset phase control: A novel concept for absolute optical frequency measurement and ultrashort pulse generation," *Appl. Phys. B*, vol. 69, pp. 327-332, 1999.

[66] D. J. Jones, S. A. Diddams, J. K. Ranka, A. Stentz, R. S. Windeler, J. L. Hall, and S. T. Cundiff, "Carrier-envelope phase control of femtosecond

mode-locked lasers and direct optical frequency synthesis," *Science*, vol. 288, pp. 635-639, Apr 2000.

[67] A. Apolonski, A. Poppe, G. Tempea, C. Spielmann, T. Udem, R. Holzwarth, T. W. Hansch, and F. Krausz, "Controlling the phase evolution of few-cycle light pulses," *Physical Review Letters*, vol. 85, pp. 740-743, Jul 2000.

[68] W. Boutu, S. Haessler, H. Merdji, P. Breger, G. Waters, M. Stankiewicz, L. J. Frasinski, R. Taieb, J. Caillat, A. Maquet, P. Monchicourt, B. Carre, and P. Salieres, "Coherent control of attosecond emission from aligned molecules," *Nature Physics*, vol. 4, pp. 545-549, Jul 2008.

[69] S. Baker, J. S. Robinson, C. A. Haworth, H. Teng, R. A. Smith, C. C. Chirila, M. Lein, J. W. G. Tisch, and J. P. Marangos, "Probing proton dynamics in molecules on an attosecond time scale," *Science*, vol. 312, pp. 424-427, 2006.

[70] B. E. A. Saleh and T. M. Carl, *Fundamentals of Photonics*: John Wiley & Sons, Inc., 2001.

[71] B. L. Henke, E. M. Gullikson, and J. C. Davis, "X-ray interactions - photoabsorption, scattering, transmission and reflection at E=50-30,000 eV, Z=1-92," *Atomic Data and Nuclear Data Tables*, vol. 54, pp. 181-342, Jul 1993.

[72] M. Sanchez del Rio and R. J. Dejus, "XOP 2.1 - a new version of the X-ray optics software toolkit," *AIP Conference Proceedings*, pp. 784-787, 2004.

[73] A. Flettner, J. Gunther, M. B. Mason, U. Weichmann, R. Duren, and G. Gerber, "High harmonic generation at 1 kHz repetition rate with a pulsed valve," *Applied Physics B-Lasers and Optics*, vol. 73, pp. 129-132, Aug 2001.

[74] A. Heinrich, M. Bruck, C. P. Hauri, W. Kornelis, J. W. G. Tisch, J. Biegert, and U. Keller, "Gas target for efficient high-harmonic generation," in *International Quantum Electronics Conference (IQEC)*, San Francisco, CA,, 2004.

[75] "Solve magnetic interference problems with magnetic shields," http://www.magnetic-shield.com/faq/interference.html, Magnetic Shield Corp.

[76] "Submicron Bandpass Foil Filters," http://www.luxel.com/standard_filters.html, Luxel Corporation.

[77] F. R. Powell, P. W. Vedder, J. F. Lindblom, and S. F. Powell, "Thin-film filter performance for extreme ultraviolet and X-ray applications," in *Conf on X-Ray/Euv Optics for Astronomy and Microscopy*, San Diego, Ca, 1989, pp. 614-624.

[78] R. Lopez-Martens, K. Varju, P. Johnsson, J. Mauritsson, Y. Mairesse, P. Salieres, M. B. Gaarde, K. J. Schafer, A. Persson, S. Svanberg, C. G. Wahlstrom, and A. L'Huillier, "Amplitude and phase control of attosecond light pulses," *Physical Review Letters,* vol. 94, p. 033001, Jan 28 2005.

[79] J. H. Weaver and H. P. R. Frederikse, "Optical Properties of Selected Elements," in *Handbook of chenistry and physics*. vol. 84, D. R. Lide, Ed.: CRC Press, 2004.

[80] J. M. Schins, P. Breger, P. Agostini, R. C. Constantinescu, H. G. Muller, A. Bouhal, G. Grillon, A. Antonetti, and A. Mysyrowicz, "Cross-correlation measurements of femtosecond extreme-ultraviolet high-order harmonics," *Journal of the Optical Society of America B-Optical Physics,* vol. 13, pp. 197-200, Jan 1996.

[81] K. Varju, P. Johnsson, R. Lopez-Martens, T. Remetter, E. Gustafsson, J. Mauritsson, M. B. Gaarde, K. J. Schafer, C. Erny, I. Sola, A. Zair, E. Constant, E. Cormier, E. Mevel, and A. L'Huillier, "Experimental

studies of attosecond pulse trains," *Laser Physics,* vol. 15, pp. 888-898, Jun 2005.

[82] *Einfuehrung in die Hoch- und Ultrahochvakuum-Erzeugung*: Pfeiffer Vacuum GmbH.

[83] E. Shiles, T. Sasaki, M. Inokuti, and D. Y. Smith, "Self-Consistency and Sum-Rule Tests in the Kramers-Kronig Analysis of Optical-Data - Applications to Aluminum," *Physical Review B,* vol. 22, pp. 1612-1628, 1980.

[84] A. Heinrich, W. Kornelis, M. P. Anscombe, C. P. Hauri, P. Schlup, J. Biegert, and U. Keller, "Enhanced VUV-assisted high harmonic generation," *J. Phys. B: At. Mol. Opt. Phys.,* vol. 39, pp. S275-S281, 2005.

[85] G. W. F. Drake and W. C. Martin, "Ionization energies and quantum electrodynamic effects in the lower 1sns and 1snp levels of neutral helium (He-4 I)," *Canadian Journal of Physics,* vol. 76, pp. 679-698, Sep 1998.

Appendix

A.1. Atomic Energy Levels for He I

Following data has been obtained from [85]. Only s and p levels are shown. The total angular quantum momentum number J is not listed.

Configuration	Level [eV]
1s2	0.0
1s.2s	19.8196134
1s.2s	20.6157736
1s.2p	20.9640857
	20.9640951
	20.9642176
1s.2p	21.2180214
1s.3s	22.7184651
1s.3s	22.9203161
1s.3p	23.0070718
	23.0070745
	23.0071081
1s.3p	23.0870172
1s.4s	23.5939573
1s.4s	23.6735693
1s.4p	23.7078898
	23.7078909
	23.7079046
1s.4p	23.7420687
1s.5s	23.9719700
1s.5s	24.0112136
1s.5p	24.0282236
	24.0282242
	24.0282311
1s.6s	24.1689968

1s.6s	24.1911589
1s.6p	24.2008140
	24.2008143
	24.2008183
1s.6p	24.2110012
1s.7s	24.2845646
1s.7s	24.2982845
1s.7p	24.3042878
	24.3042880
	24.3042904
1s.7p	24.3107067
1s.8s	24.3581027
1s.8s	24.3671786
1s.8p	24.3711644
	24.3711646
	24.3711662
1s.8p	24.3754653
1s.9s	24.4077739
1s.9s	24.4140866
1s.9p	24.4168674
	24.4168675
	24.4168687
1s.9d	24.4193341
	24.4193341
	24.4193343
1s.9d	24.4193617
1s.9p	24.4198879
1s.10s	24.4428920
1s.10s	24.4474586
1s.10p	24.4494754
	24.4494755
	24.4494763
1s.10p	24.4516770
He II (2S<1/2>)	24.5873876

List of Abbreviations

ABS	active beam stabilization
APT	attosecond pulse train
BS	barrier suppression
BSI	barrier suppression intensity
CCD	charged coupled device
CEP	carrier-envelope offset phase
CPA	chirped pulse amplification
FWHM	full width at half maxium
GDD	group delay dispersion
HHG	high harmonic generation
IR	infrared
MCP	microchannel plate
QPI	first order quantum path interferences
RABBITT	reconstruction of attosecond beating by interference of two photon transitions
ROC	radius of curvature
SFA	strong field approximation
TDSE	time-dependent Schrödinger equation
Ti:sapphire	titanium doped sapphire
TOF	time-of-flight
UV	ultraviolet
XUV	extreme ultraviolet

Danksagung

An dieser Stelle möchte ich mich bei allen Personen ganz herzlich bedanken die zum Erfolg dieser Doktorarbeit beigetragen haben und mich während dieser Zeit unterstützt haben. Ohne euer Zutun wären die durchgeführten Projekte nicht möglich gewesen. Im Besonderen:

Ursi, dafür, dass ich meine Doktorarbeit in deiner Gruppe machen durfte. Speziell danke ich für deine Unterstützung des attoline Projekts, wo ich mich verwirklichen durfte.

Prof. Dr. Ian Walmsley, dafür dass Sie das Koreferat übernommen haben und für die Zusammenarbeit im QPI Projekt.

Lukas, für die vielen kompetenten Diskussionen und für die Freiheit, die du uns gegeben hast.

Florian, wir waren ein super Team! Nur dadurch sind die Ergebnisse überhaupt möglich geworden.

Arne, was ich von dir zu Beginn meiner Diss. gelernt habe, hat mir bis zum Ende geholfen.

Annalisa, Laser schrauben war schon cool.

Amelle, nach langer Suche haben wir die QPI doch noch gefunden.

Erik, nach langen two-color Messungen noch schnell Regale im Labor absägen...

Petrissa, Pulse messen, dann Mittagessen im F8 und danach noch zum Baumarkt.

Marcel, danke für viele interessante Stunden im Ingenieurbüro. War super, dass ich dich bezüglich CATIA immer belästigen durfte. An dieser Stelle auch vielen Dank an Walter und Martina.

Hansruedi, vielen Dank für die Aluminium Filter, die du mit deinen Lehrlingen für uns in allen denkbaren Varianten erstellt hast.

Andi, vielen Dank für die Unterstützung von dir und deinem gesamten Team der Werkstatt. Erst durch euch sind die Pläne der attoline Wirklichkeit geworden.

Danksagung

Stefan, deine guten Tips und Ratschläge begleiten mich schon mehr als 10 Jahre und haben mich überhaupt erst zur Physik geführt. Vielen Dank dafür.

Des weiteren danke ich allen restlichen aktuellen und ehemaligen Mitgliedern der sub10 Gruppe:

Anoush Aghajani, Marcel Anscombe, Claudio Cirelli, Christian Erny, Christoph Hauri, Clemens Heese, Reto Locher, Konstantinous Moutzouris, Adrian Pfeiffer, Thomas Remetter, Philip Schlup, Mathias Smolarski, Matthias Weger.

Ich möchte mich auch ganz herzliche bei allen anderen aktuellen und ehemaligen ULP-Mitgliedern für die gute Arbeitsatmosphäre und Stimmung in der Gruppe bedanken:

Yohan Barbarin, Cyrill Bär, Aude-Reine Bellancourt, Anna Engqvist, Birgit Gallmann, Anastassia Gosteva, Matthias Golling, Rachel Grange, Shigeki Hashimoto, Oliver Heckel, Martin Hoffmann, Edith Innerhofer, Christian Kränkel, Dietrich Kühlke, Valeria Liverini, Deran Maas, Sergio Marchese, Paolo Navaretti, Andreas Oehler, Wolfgnang Pallmann, Selina Pekarek, Benjamin Rudin, Andreas Rutz, Clara Saranceno, Adrian Schlatter, Sandra Schmid, Silke Schön, Oliver Sieber, Max Stumpf, Thomas Südmeyer, Heiko Unold, Valentin Wittwer und Simon Zeller.

Ein besonderer Dank gilt auch meinen externen Kollaboratoren:

Bei den QPI: Thierry Auguste, Jean-Pascal Caumes, Eric Cormier, Antoine Monmayrant, Pascal Salieres, Adam Wyatt.

Zusätzlich bei den two-color Messungen: Markus Dahlström, Mette Gaarde, Anne L'Huillier, Aurelie Jullien, Johan Mauritsson, Hamed Merdji, Thierry Ruchon and Jennifer Tate.

Schliesslich gilt ein ganz besonderer Dank meinen Eltern, die mich mein ganzes Leben bei meinen Plänen und Vorhaben stets unterstützt haben. Und natürlich meiner Ehefrau, Madlaina, die für meine Physik-Geschichten stets ein offenes Ohr hatte.

Zürich, im Dezember 2009,
Mirko Holler.

Die VDM Verlagsservicegesellschaft sucht für wissenschaftliche Verlage abgeschlossene und herausragende

Dissertationen, Habilitationen, Diplomarbeiten, Master Theses, Magisterarbeiten usw.

für die kostenlose Publikation als Fachbuch.

Sie verfügen über eine Arbeit, die hohen inhaltlichen und formalen Ansprüchen genügt, und haben Interesse an einer honorarvergüteten Publikation?

Dann senden Sie bitte erste Informationen über sich und Ihre Arbeit per Email an *info@vdm-vsg.de*.

Sie erhalten kurzfristig unser Feedback!

VDM Verlagsservicegesellschaft mbH
Dudweiler Landstr. 99
D - 66123 Saarbrücken

Telefon +49 681 3720 174
Fax +49 681 3720 1749

www.vdm-vsg.de

Die VDM Verlagsservicegesellschaft mbH vertritt

Printed by Books on Demand GmbH, Norderstedt / Germany